Introduction to Entomology

Introduction to Entomology

Claudia Sherwood

MURPHY & MOORE
www.murphy-moorepublishing.com

Murphy & Moore Publishing,
1 Rockefeller Plaza,
New York City, NY 10020, USA

Visit us on the World Wide Web at:
www.murphy-moorepublishing.com

ISBN: 978-1-63987-330-2 (Hardback)

Cataloging-in-Publication Data

Introduction to entomology / Claudia Sherwood.
p. cm.
Includes bibliographical references and index.
ISBN 978-1-63987-330-2
1. Entomology. 2. Insects. 3. Zoology. I. Sherwood, Claudia.
QL463 .I58 2022
595.7--dc23

TABLE OF CONTENTS

PREFACE

This book has been written, keeping in view that students want more practical information. Thus, my aim has been to make it as comprehensive as possible for the readers. I would like to extend my thanks to my family and co-workers for their knowledge, support and encouragement all along.

Entomology is the branch of zoology, which deals with the scientific study of insects. It focuses on how insects interact with the environment and other species. It plays a crucial role in the assessment of environmental quality and study of biodiversity. There are many sub-specialties of entomology such as coleopterology, dipterology, melittology, myrmecology, lepidopterology, odonatology, etc. Medical entomology is concerned with the medical research of insects and their impact on public health. It also studies plant damage and diseases that are caused by insects, along with effectiveness and development of pesticides. Economic entomology is associated with the study of harmful and beneficial impact of insects on humans, animals and crops. This book aims to shed light on some of the unexplored aspects of entomology. It is a valuable compilation of topics, ranging from the basic to the most complex theories and principles in this field. This textbook will provide comprehensive knowledge to the readers.

A brief description of the chapters is provided below for further understanding:

Chapter - What is Entomology?

Entomology is the branch of zoology that deals with the study of insects. This discipline is used as environmental indicators, for archaeology and in fruit fly research. This is an introductory chapter which will briefly introduce about the field of entomology.

Chapter - Classification of Insects

The species of insects can be classified into different groups. Some of them are Coleopteran, Diptera, Hymenoptera, Lepidoptera, Blattodea, Hemiptera, etc. This chapter closely examines these different classified groups of insects to provide an extensive understanding of the subject.

Chapter - External Anatomy

External anatomy of insects comprises of three major body regions - head, thorax and abdomen. Head has five to seven fused segments with eyes, antennae and mouthparts. Thorax can be divided into three segments and abdomen comprises eleven or fewer segments. This chapter discusses the external anatomy of insects in detail.

Chapter - Internal Anatomy

The internal anatomy of insects consists of the nervous system, digestive system, circulatory system, respiratory system, muscular system, endocrine system and reproductive system. The topics elaborated in this chapter will help in gaining a better perspective of internal anatomy of insects.

Chapter - Evolution of Insects

Evolution of color vision, evolution relationships, evolution of circadian clocks, evolution of eusocial insects, evolution of osmoregulatory capacity, evolution of tracheal systems, etc. are studied under the domain of evolution of insects. This chapter has been carefully written to provide an easy understanding of the subject of evolution of insects.

Chapter - Conservation of Insects

The process of caring and preserving insects through collection, preparation, storage, examination, documentation, research and treatment is referred to as insect conservation. Its psychology and practices are also studied within it. This chapter sheds light on the subject of conservation of insects for an in-depth understanding of it.

Chapter - Insects Interactions

The relationship between insects and humans fits into all three categories of a biotic relationship i.e. parasitism, commensalism and mutualism. Negative impacts of insects include transmitting diseases, damage to structures, and positive effects consists of pollination, creating valuable goods like silk, lacquer, dyes, etc. All the aspects of this relationship between insects and humans have been carefully analyzed in this chapter.

Claudia Sherwood

What is Entomology?

1 CHAPTER

Entomology is the branch of zoology that deals with the study of insects. This discipline is used as environmental indicators, for archaeology and in fruit fly research. This is an introductory chapter which will briefly introduce about the field of entomology.

Entomology is the scientific study of insects. Insects are arthropods (phylum Arthropoda) belonging to the Class Insecta. With around 925,000 described species, insects comprise the most numerous and diverse group of animals, representing more than half (about 57 percent) of all identified animal species.

Part of a large beetle collection.

Insects have many kinds of interactions with humans and other forms of life on earth, including key roles in food webs and nutrient recycling, in symbiotic relationships with flowering plants as pollinators, as producers of such human products like silk, honey, and shellac, and as agricultural pests and vectors of disease. Given these factors and the sheer numbers of insects—combined with a human nature endowed with curiosity and creativity and an often viewed role as stewards of nature—it is not surprising that entomology is an important specialty within biology.

Though technically incorrect, the definition of entomology is sometimes widened to include the study of terrestrial invertebrates in other arthropod groups or other phyla, such as arachnids, myriapods, earthworms, and slugs.

Applied Entomology

Many entomologists are employed in the study of insects that are directly beneficial or harmful to humans. Beneficial insects include those that are pollinators of agricultural crops and ornamental flowers and those imported or otherwise vital as biological control agents. The study of beneficial insects is primarily focused on their ecology and life habits, the primary concern being the understanding of how to raise them and make them more productive, or protect them from human disturbance if they are native species such as wild bees.

Conversely, much of the study of insects (and related arthropods) that directly harm human beings or their domestic animals, called medical entomology and veterinary entomology, is focused on their physiology, with the goal of developing insect controls that are effective, while minimizing undesirable side effects. For instance, many types of insecticides have been developed that target unique aspects of insect physiology and are thus considered harmless to other kinds of animals. A risk to this approach is that insecticides can also kill beneficial insects. Considerable recent effort has gone into finding biological controls that are species-specific, such as species-specific parasites and diseases, as well as genetic controls, such as the introduction of sterile insects into a population. The combination of taking into account all aspects of insect biology, available control measures, economics, and environmental considerations is known as integrated pest management.

A few insects, chiefly blood-sucking Diptera, are vectors for a wide range of deadly diseases. Mosquitoes are especially important disease vectors, with the genus Anopheles the principle vector of malaria, Aedes aegypti the main vector of yellow fever and dengue, and other Aedes spp. carrying the causal agents of various types of encephalitis. Other well-known vectors include the tsetse fly (genus Glossina transmits protozoan agents of African sleeping sickness), black flies (spread the parasitic roundworm Onchocerca volvulus, cause of onchoceriasis), and sand flies (genus Phlebotomus are vectors of bacteria that causes Carrion's disease, and sand flies are also agent of protozoans Leishmania spp. that cause Leishmaniasis).

Forensic entomology specializes in the study of insect ecology for use in the legal system, as knowledge of insect behavior can yield useful information about crimes. For example, the approximate time of death or whether or not a victim was alive during a fire may be determined by using facts such as at what stage of life is an insect found at the scene.

Taxonomic Specialization

Given the vast number and diversity of insects, many entomologists specialize in a single order or even a family of insects. A number of these subspecialties are given their own informal names, typically (but not always) derived from the scientific name of the group:

- Apiology (or melittology): (Study of) bees,

- Coleopterology: Beetles,
- Dipterology: Flies,
- Heteropterology: True bugs,
- Lepidopterology: Moths and butterflies,
- Myrmecology: Ants,
- Orthopterology: Grasshoppers, crickets, etc..
- Trichopterology: Caddis flies.

Identification of Insects

Insects other than Lepidoptera are typically identifiable only through the use of identification keys and monographs. Because the class Insecta contains a very large number of species, and the characters separating them are unfamiliar and often subtle (or invisible without a microscope), this is often very difficult even for a specialist.

Insect identification is an increasingly common hobby, with butterflies and dragonflies being the most popular.

USES OF ENTOMOLOGY

We discover new uses for studies in entomology all the time and there are already several established disciplines where insects and insect remains are a powerful type of evidence.

As Environmental Indicators

Insects are a vital part of forest biodiversity and as they are particularly sensitive to changes in the climate and the patterns and seasons of regional flora cover, entomology provides evidence for general forest health as well changes in the cover. The arrival of a new pest or the sudden decline of a well-established native species can indicate many things, such as the effects of deforestation, or a change in the types of arboreal cover, CO_2 concentration, persistent ecological issues such as drought or flooding. They can tell us much about the types of trees, shrubs and flowers that grow there (especially pollinating insects such as butterflies) and any changes.

The Arctic Circle has a very delicate ecosystem and this is one area where researchers in many fields rely heavily on entomology, particularly of the relationship between insect groups, between insects and the environment, between insects and other animal species; evidence from population numbers and density is often used to examine specific environmental indicators.

For Archaeology/Anthropology

Similarly, and as archaeology moves more into the realm of environmental science, insect population and remains are becoming more important in the interpretation of archaeological sites and landscapes. Humans have had an enormous impact on the environment of the past, and we are just beginning to understand how geo-engineering, how hunter-gatherer societies and even the Neolithic Agricultural Revolution impacted local ecologies. We are only now beginning to understand how insect remains are useful, and indeed which insect species are most useful in which areas. In a study of western Greenland for example, researchers examined extant insect remains for indicators of personal hygiene.

This isn't all they can tell us though. When a virgin landscape is altered to make way for farmland, there is likely to be a change to the ecosystem that will include a change of both the pollen distribution pattern and insect remains - especially where a crop favourable to native insect species replaces a crop that was unfavourable, or vice versa. In these cases, there will be a clear and distinct pattern change in the archaeological record.

Forensic Entomology

It may surprise people to learn that insect species can sometimes play a part in criminal investigation and even be included as part of the evidence. We can learn much about a body's season of death and for how long it may have been exposed to the elements before discovery or being buried. This is one of the most critical areas in which entomology contributes to our understanding of human remains - in archaeology and in criminal proceedings.

Insects are attracted to decomposing bodies and may begin feeding off it or laying eggs in it. By understanding the lifecycle process of any insect remains found, investigators can identify how long a body has been dead, where it has been storied and a variety of other facts. Most interestingly, it seems that studying insects as a method of working out the perpetrator of a murder goes back to China in the 13th century when suspects were told to lay their sickles on the ground to see which attracted the most flies to the small amounts of blood that remained on the blade. Using insects as evidence in a modern criminal trial began in 1935 when a body in the UK showed a number of blowflies. The subsequent evidence attained from the blowflies informed investigators how long the woman had been dead and led to a conviction of the guilty party.

Medical Entomology

This is the area concerned with medical research and how insects are a matter of public health. This area also includes plant disease and damage caused by insects to crops too. That means researchers in Medical Entomology are concerned with urban and rural pests as well as the diseases that some insects carry. The scope of medical entomology will also cover research into the effectiveness of, and development of new, pesticides.

Medical entomology arguably began in the early 20th century when it was discovered that mosquitoes were responsible for the transmission of Yellow Fever. Medical entomology plays a large part of the US military research as the major bodies examined and combated the effects of major diseases in all wars the country has participated in - most recently in Iraq and Afghanistan. Though entomology is the study of insects, medical entomology has a broader scope in that it incorporates other arthropods that may affect human health - this means arachnids such as spiders, mites, ticks and also come under the scope of a medical entomology researcher.

Fruit Fly Research

The humble fruit fly deserves a special mention because these small insects have given us so much information. We have learnt much about evolutionary biology because many generations breed in a very short space of time. We have information on genetic mutation, genetic drift and other elements of evolution vital to research in other areas, particularly human genetics. Selective cross-breeding for favourable traits may be carried out on a large scale in a very short space of time. Though they have been studied for at least a century, only recently have researchers made great leaps in discoveries.

INSECT BIODIVERSITY

Insect, (class Insecta or Hexapoda) is any member of the largest class of the phylum Arthropoda, which is itself the largest of the animal phyla. Insects have segmented bodies, jointed legs, and external skeletons (exoskeletons). Insects are distinguished from other arthropods by their body, which is divided into three major regions: (1) the head, which bears the mouthparts, eyes, and a pair of antennae, (2) the three-segmented thorax, which usually has three pairs of legs (hence “Hexapoda”) in adults and usually one or two pairs of wings, and (3) the many-segmented abdomen, which contains the digestive, excretory, and reproductive organs.

In a popular sense, “insect” usually refers to familiar pests or disease carriers, such as bedbugs, houseflies, clothes moths, Japanese beetles, aphids, mosquitoes, fleas, horseflies, and hornets, or to conspicuous groups, such as butterflies, moths, and beetles. Many insects, however, are beneficial from a human viewpoint; they pollinate plants, produce useful substances, control pest insects, act as scavengers, and serve as food for other animals. Furthermore, insects are valuable objects of study in elucidating many aspects of biology and ecology. Much of the scientific knowledge of genetics has been gained from fruit fly experiments and of population biology from flour beetle studies. Insects are often used in investigations of hormonal action, nerve and sense organ function, and many other physiological processes. Insects are also used as environmental quality indicators to assess water quality and soil contamination and are the basis of many studies of biodiversity.

Paper wasp.

European hornet (Vespa crabro).

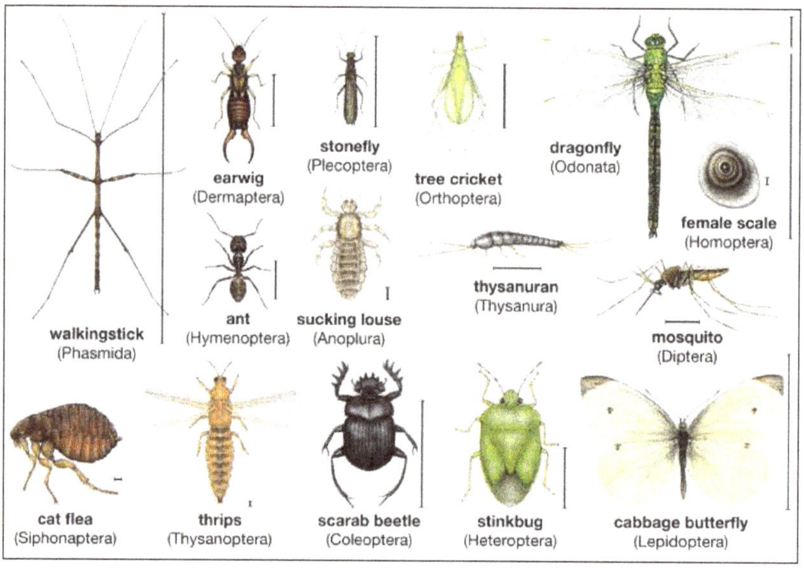

Insect diversity.

General Features

In numbers of species and individuals and in adaptability and wide distribution, insects are perhaps the most eminently successful group of all animals. They dominate the present-day land fauna with about 1 million described species. This represents about three-fourths of all described animal species. Entomologists estimate the actual number of living insect species could be as high as 5 million to 10 million. The orders that contain the greatest numbers of species are Coleoptera (beetles), Lepidoptera (butterflies and moths), Hymenoptera (ants, bees, wasps), and Diptera (true flies).

Eastern tailed blue butterfly (Everes comyntas; also called Cupido comyntas).

Bombardier beetle (Brachinus).

Appearance and Habits

The majority of insects are small, usually less than 6 mm (0.2 inch) long, although the range in size is wide. Some of the feather-winged beetles and parasitic wasps are almost microscopic, while some tropical forms such as the hercules beetles, African goliath beetles, certain Australian stick insects, and the wingspan of the hercules moth can be as large as 27 cm (10.6 inches).

African goliath beetle (Goliathus giganteus).

Walkingstick (Phasmatidae).

In many species the difference in body structure between the sexes is pronounced, and knowledge of one sex may give few clues to the appearance of the other sex. In some, such as the twisted-wing insects (Strepsiptera), the female is a mere inactive bag of eggs, and the winged male is one of the most active insects known. Modes of reproduction are quite diverse, and reproductive capacity is generally high. Some insects, such as the mayflies, feed only in the immature or larval stage and go without food during an extremely short adult life. Among social insects, queen termites may live for up to 50 years, whereas some adult mayflies live less than two hours.

Female mayfly (Ephemera danica).

Some insects advertise their presence to the other sex by flashing lights, and many imitate other insects in colour and form and thus avoid or minimize attack by predators that feed by day and find their prey visually, as do birds, lizards, and other insects.

Behaviour is diverse, from the almost inert parasitic forms, whose larvae lie in the nutrient bloodstreams of their hosts and feed by absorption, to dragonflies that pursue victims in the air, tiger beetles that outrun prey on land, and predaceous water beetles that outswim prey in water.

North American firefly (Photinus).

In some cases the adult insects make elaborate preparations for the young, in others the mother alone defends or feeds her young, and in still others the young are supported by complex insect societies. Some colonies of social insects, such as tropical termites and ants, may reach populations of millions of inhabitants.

Distribution and Abundance

Scientists familiar with insects realize the difficulty in attempting to estimate individual numbers of insects beyond areas of a few acres or a few square miles in extent. Figures soon become so large as to be incomprehensible. The large populations and great variety of insects are related to their small size, high rates of reproduction, and abundance of suitable food supplies. Insects abound in the tropics, both in numbers of different kinds and in numbers of individuals.

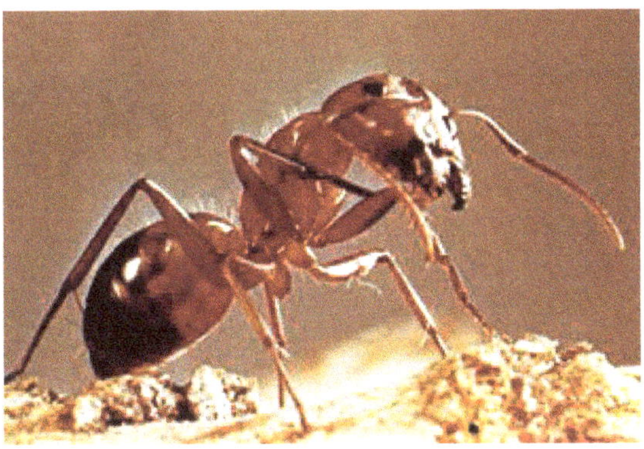

Carpenter ant (Camponotus).

If the insects (including the young and adults of all forms) are counted on a square yard (0.84 square metre) of rich moist surface soil, 500 are found easily and 2,000 are not unusual in soil samples in the north temperate zone. This amounts to roughly 4 million insects on one moist acre (0.41 hectare). In such an area only an occasional butterfly, bumblebee, or large beetle, supergiants among insects, probably would be noticed. Only a few thousand species, those that attack people's crops, herds, and products and those that carry disease, interfere with human life seriously enough to require control measures.

Ladybug.

Insects are adapted to every land and freshwater habitat where food is available, from deserts to jungles, from glacial fields and cold mountain streams to stagnant, lowland ponds and hot springs. Many live in brackish water up to $^{1}/_{10}$ the salinity of seawater, a few live on the surface of seawater, and some fly larvae can live in pools of crude petroleum, where they eat other insects that fall in.

Importance

Role in Nature

Codling moth larva (Cydia pomonella) parasitizing an apple.

Insects play many important roles in nature. They aid bacteria, fungi, and other organisms in the decomposition of organic matter and in soil formation. The decay of carrion, for example, brought about mainly by bacteria, is accelerated by the maggots of flesh flies and blowflies. The activities of these larvae, which distribute and consume bacteria, are followed by those of moths and beetles, which break down hair and feathers. Insects and flowers have evolved together. Many plants depend on insects for pollination. Some insects are predators of others.

Commercial Significance

Certain insects provide sources of commercially important products such as honey, silk, wax, dyes, or pigments, all of which can be of direct benefit to humans. Because they feed on many types of organic matter, insects can cause considerable agricultural damage. Insect pests devour crops of food or timber, either in the field or in storage, and convey infective microorganisms to crops, farm animals, and humans. The technology for combatting such pests constitutes the applied sciences of agricultural and forest entomology, stored product entomology, medical and veterinary entomology, and urban entomology.

Insects as a Source of Raw Materials

For primitive peoples who gathered food, insects were a significant food source. Grasshopper plagues, termite swarms, large palm weevil grubs, and other insects are still sources of protein in some countries. The dry scaly excreta of coccids (Homoptera) on tamarisk or larch trees is the source of manna in the Sinai Desert. Coccids were once the source of the crimson dye kermes. The cochineal, or carmine, from Dactylopius scale insects found on Mexican cacti, was used for dying cloth by the Aztecs and is used today as a dye in foods, cosmetics, drugs, and textiles. Several insect waxes are used commercially, especially beeswax and lac wax. The resinous product of the lac insect Kerria lacca (Homoptera), which is cultured for this purpose, is the source of commercial shellac.

A silkworm spinning a cocoon.

Two of the most important domesticated insects are the silkworm (Lepidoptera) and the honeybee (Hymenoptera). Some coarse silks are produced from the cocoons of large wild silkworm species. Most commercial silks, however, come from the silkworm

Bombyx mori. This insect is unknown in the wild state and exists only in culture. It was domesticated in China thousands of years ago, and selective breeding, notably in China and Japan, has produced many specialized strains. The honeybee is a close relative of existing wild bees. In the Middle Ages, honey was Europe's most important sweetener, and both beeswax and honey are still articles of commerce. However, the major importance of honeybees lies in their pollination of fruit trees and other crops.

Insect Damage to Commercial Products

When insects that break down dead trees invade structural timbers in buildings, they become pests. This is true of insects such as dermestid beetles and various tineid moths that ecologically are latecomers to carcasses and are capable of breaking down the keratin in hair and feathers. When these insects invade skins, furs, and wool garments or carpets, they can become problems for humans.

In many hot, dry climates, as in North Africa or the plains of India, ripened grain in the fields is invaded by certain beetles and moths. When the grain is harvested, these insects thrive in the grain stores. They can be carried throughout the world in commerce and have become universal pests of stored grain, dried fruit, tobacco, and other products. Quarantine and disinfestation methods are used to control importation of such insects from grain-exporting countries.

Agricultural Significance

Ecological Factors

Spotted cucumber beetle (Diabrotica undecimpunctata).

Many insects are plant feeders, and, when the plants are of agricultural importance, humans are often forced to compete with these insects. Populations of insects are limited by such factors as unfavourable weather, predators and parasites, and viral, bacterial, and fungal diseases, as well as many other factors that operate to make insect populations stable. Agricultural methods that encourage the planting of ever larger areas to single crops, which provides virtually unlimited food resources, has removed some of these regulating factors and allowed the rate of population growth of insects that attack those crops to increase. This increases the probability of great infestations of certain insect pests. Many natural forests, which form similar giant monocultures, always seem to have been subject to periodic outbreaks of destructive insects.

In some agricultural monocultures, nonnative insect pests have been accidentally introduced along with a crop but without also bringing along its full range of natural enemies. This has occurred in the United States with the oystershell scale (Lepidosaphes ulmi) of apple and other fruit trees, the cottony cushion scale (Icerya purchasi) of citrus, the European corn borer (Pyrausta nubilalis; also called Ostrinia nubilalis), and others. The Colorado potato beetle (Leptinotarsa decemlineata), which caused appalling destruction to the cultivated potato in the United States beginning about 1840, was a native insect of semidesert country. The beetle, which fed on the buffalo burr plant, adapted itself to a newly introduced and abundant diet of potatoes and thus escaped from all previous controlling factors. Similar situations often have been controlled by determining the major predators or parasites of an alien insect pest in its country of origin and introducing them as control agents. A classic example is the cottony cushion scale, which threatened the California citrus industry in 1886. A predatory ladybird beetle, the vedalia beetle (Rodolia cardinalis), was introduced from Australia, and within a year or two the scale insect had virtually disappeared. The success was repeated in every country where the scale insect had become established without its predators. In eastern Canada in in the early 1940s the European spruce sawfly (Gilpinia hercyniae), which had caused immense damage, was completely controlled by the spontaneous appearance of a viral disease, perhaps unknowingly introduced from Europe. This event led to increased interest in using insect diseases as potential means of managing pest populations.

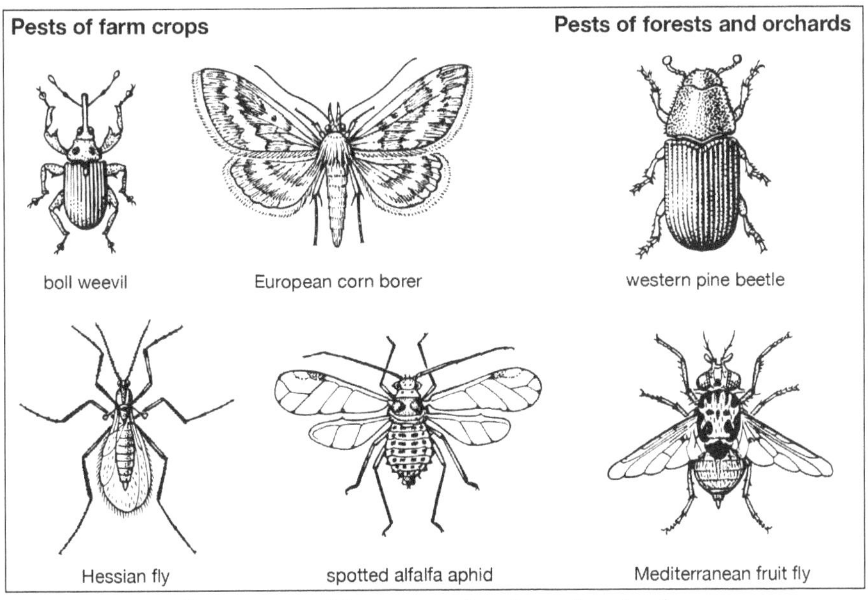

Insect pests of farm crops, forests, and orchards.

The boll weevil (Anthonomus grandis), European corn borer (Pyrausta nubilalis, or Ostrinia nubilalis), Hessian fly (Mayetiola destructor), and spotted alfalfa aphid (Therioaphis maculata) are important pests of farm crops. The western pine beetle (Dendroctonus brevicomis) is an insect pest found in forests, while the Mediterranean fruit fly (Ceratitis capitata) is an example of an orchard pest.

Cottony-cushion scales (Icerya purchasi, magnified).

Damage to Growing Crops

Insects are responsible for two major kinds of damage to growing crops. First is direct injury done to the plant by the feeding insect, which eats leaves or burrows in stems, fruit, or roots. There are hundreds of pest species of this type, both in larvae and adults, among orthopterans, homopterans, heteropterans, coleopterans, lepidopterans, and dipterans. The second type is indirect damage in which the insect itself does little or no harm but transmits a bacterial, viral, or fungal infection into a crop. Examples include the viral diseases of sugar beets and potatoes, carried from plant to plant by aphids.

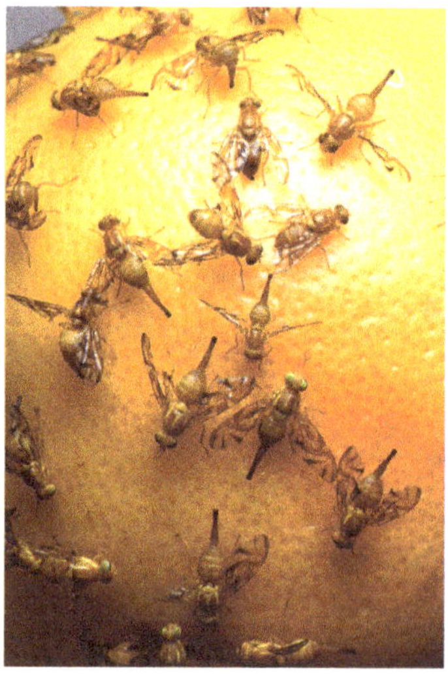

Mexican fruit flies (Anastrepha ludens), an invasive species, feeding on a citrus fruit.

Although most insects grow and multiply in the crop they damage, certain grasshoppers are well-known exceptions. They can exist in a relatively harmless solitary phase for a number of years, during which time their numbers may increase. They then enter a gregarious phase, forming gigantic migratory swarms, which are transported by winds or flight for hundreds or thousands of miles. These swarms may completely destroy crops in an invaded region. The desert locust (Schistocerca gregaria) and migratory locust (Locusta migratoria) are two examples of this type of life cycle.

Medical Significance

Insect damage to humans and livestock also may be direct or indirect. Direct human injury by insect stings and bites is of relatively minor importance, although swarms of biting flies and mosquitoes often make life almost intolerable, as do biting midges (sand flies) and salt-marsh mosquitoes. Persistent irritation by biting flies can cause deterioration in the health of cattle. Some blowflies, in addition to depositing their eggs in carcasses, also invade the tissue of living animals including humans, a condition known as myiasis. An example of an insect that causes this condition is the screwworm fly (Cochliomyia) of the southern United States and Central America. In many parts of the world, various blowflies infest the fleece and skin of sheep. This infestation, called sheep-strike, causes severe economic damage.

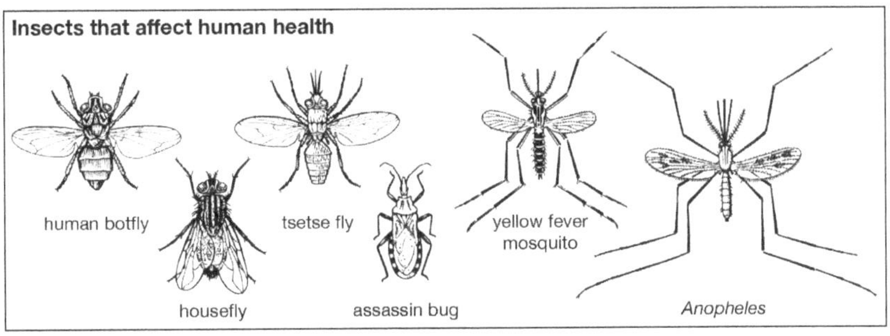

Insects that affect human health.

The human botfly (Dermatobia hominis), housefly (Musca domestica), tsetse fly (genus Glossina), assassin bug (family Reduviidae), yellow fever mosquito, and Anopheles mosquito are examples of insects that are of medical significance to humans.

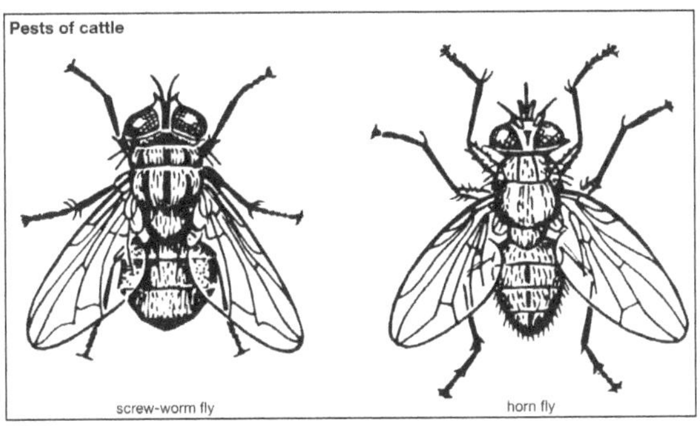

Insect pests of cattle:The screwworm and horn fly (Haematobia irritans) are insect pests of cattle.

Many major human diseases are produced by microorganisms conveyed by insects, which serve as vectors of pathogens. Malaria is caused by the protozoan Plasmodium, which spends part of its developmental cycle in Anopheles mosquitoes. Epidemic relapsing fever, caused by spirochetes, is transmitted by the louse Pediculus. Leishmaniasis, caused by the protozoan Leishmania, is carried by the sand fly Phlebotomus. Sleeping sickness in humans and a group of cattle diseases that are widespread in Africa and known as nagana are caused by protozoan trypanosomes transmitted by the bites of tsetse flies (Glossina). Under nonsanitary conditions the common housefly Musca can play an incidental role in the spread of human intestinal infections (e.g., typhoid, bacillary and amebic dysentery) by contamination of food. The tularemia bacillus can be spread by deerfly bites, the bubonic plague bacillus by fleas, and the epidemic typhus rickettsia by the louse Pediculus. Various mosquitoes spread viral diseases (e.g., several encephalitis diseases; dengue and yellow fever in humans and other animals).

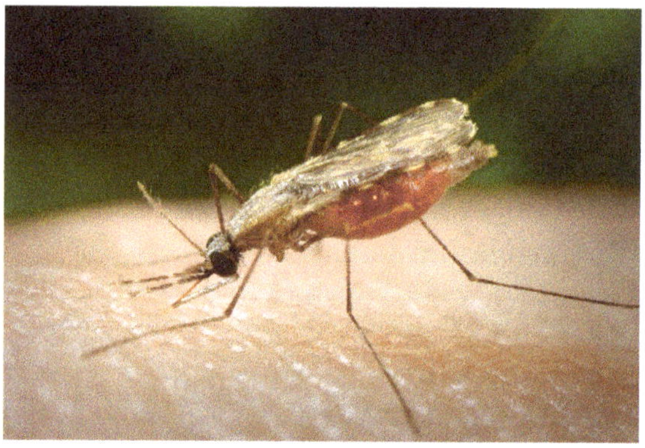

Mosquito (Anopheles minimus) feeding on a human. A. minimus is a major malaria vector in Asia.

The relationships among the various organisms are complex. Malaria, for example, has a different epidemiology in almost every country in which it occurs, with different Anopheles species responsible for its spread. These same complexities affect the spread of sleeping sickness. Some relationships are indirect. Plague, a disease of rodents transmitted by flea bites, is dangerous to humans only when heavy mortality among domestic rats forces their infected fleas to attack people, thereby causing an outbreak of plague. Typhus, tularemia, encephalitis, and yellow fever also are maintained in animal reservoirs and spread occasionally to humans.

Classification of Insects

CHAPTER 2

The species of insects can be classified into different groups. Some of them are Coleopteran, Diptera, Hymenoptera, Lepidoptera, Blattodea, Hemiptera, etc. This chapter closely examines these different classified groups of insects to provide an extensive understanding of the subject.

Insects are the largest group of animals, which constitute 75% of all living animals. There are more than one million living species in the world and many more yet to be discovered. They were the first flying creatures and still the only invertebrates that can fly. They are the most successful animals inhabiting every conceivable ecological condition.

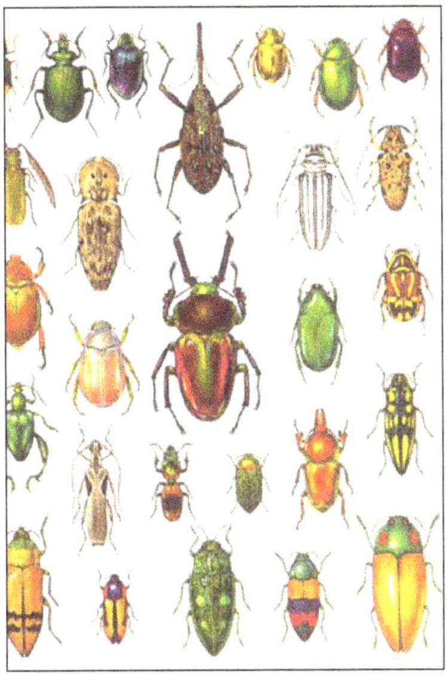

Although only 2% of the insect species are obnoxious to man, they are enough to cause heavy damage to crop, livestock and man himself. That makes them very important for human attention. Pest species are difficult to control and in spite of about a century of constant efforts, cockroaches, flies and mosquitoes are as abundant as ever. Insect bodies are extremely tough due to exoskeleton and they possess enormous muscle power. They are endowed with the capacity to survive in the most inhospitable situations.

Classification of Insects on the basis of species is as follows:

- Coleoptera,

- Diptera,
- Hymenoptera,
- Lepidoptera,
- Blattodea,
- Hemiptera.

COLEOPTERA

Beetles are a group of insects that form the order Coleoptera, in the superorder Endopterygota. Their front pair of wings are hardened into wing-cases, elytra, distinguishing them from most other insects. The Coleoptera, with about 400,000 species, is the largest of all orders, constituting almost 40% of described insects and 25% of all known animal life-forms; new species are discovered frequently. The largest of all families, the Curculionidae (weevils), with some 83,000 member species, belongs to this order. Found in almost every habitat except the sea and the polar regions, they interact with their ecosystems in several ways: beetles often feed on plants and fungi, break down animal and plant debris, and eat other invertebrates. Some species are serious agricultural pests, such as the Colorado potato beetle, while others such as Coccinellidae (ladybirds or ladybugs) eat aphids, scale insects, thrips, and other plant-sucking insects that damage crops.

Beetles typically have a particularly hard exoskeleton including the elytra, though some such as the rove beetles have very short elytra while blister beetles have softer elytra. The general anatomy of a beetle is quite uniform and typical of insects, although there are several examples of novelty, such as adaptations in water beetles which trap air bubbles under the elytra for use while diving. Beetles are endopterygotes, which means that they undergo complete metamorphosis, with a series of conspicuous and relatively abrupt changes in body structure between hatching and becoming adult after a relatively immobile pupal stage. Some, such as stag beetles, have a marked sexual dimorphism, the males possessing enormously enlarged mandibles which they use to fight other males. Many beetles are aposematic, with bright colours and patterns warning of their toxicity, while others are harmless Batesian mimics of such insects. Many beetles, including those that live in sandy places, have effective camouflage.

Beetles are prominent in human culture, from the sacred scarabs of ancient Egypt to beetlewing art and use as pets or fighting insects for entertainment and gambling. Many beetle groups are brightly and attractively coloured making them objects of collection and decorative displays. Over 300 species are used as food, mostly as larvae; species widely consumed include mealworms and rhinoceros beetle larvae. However, the major impact of beetles on human life is as agricultural, forestry, and horticultural pests. Serious pests include the boll weevil of cotton, the Colorado potato beetle, the coconut hispine beetle, and the mountain pine beetle. Most beetles, however, do not cause economic damage and many, such as the lady beetles and dung beetles are beneficial by helping to control insect pests.

Distribution and Diversity

Beetles are by far the largest order of insects: The roughly 400,000 species make up about 40% of all insect species so far described, and about 25% of all animals. A 2015 study provided four independent estimates of the total number of beetle species, giving a mean estimate of some 1.5 million with a "surprisingly narrow range" spanning all four estimates from a minimum of 0.9 to a maximum of 2.1 million beetle species. The four estimates made use of host-specificity relationships (1.5 to 1.9 million), ratios with other taxa (0.9 to 1.2 million), plant: Beetle ratios (1.2 to 1.3), and extrapolations based on body size by year of description (1.7 to 2.1 million).

Beetles are found in nearly all habitats, including freshwater and coastal habitats, wherever vegetative foliage is found, from trees and their bark to flowers, leaves, and underground near roots - even inside plants in galls, in every plant tissue, including dead or decaying ones. Tropical forest canopies have a large and diverse fauna of beetles, including Carabidae, Chrysomelidae, and Scarabaeidae.

The heaviest beetle, indeed the heaviest insect stage, is the larva of the goliath beetle, *Goliathus goliatus*, which can attain a mass of at least 115 g (4.1 oz) and a length of 11.5 cm (4.5 in). Adult male goliath beetles are the heaviest beetle in its adult stage, weighing 70–100 g (2.5–3.5 oz) and measuring up to 11 cm (4.3 in). Adult elephant beetles, *Megasoma elephas* and *Megasoma actaeon* often reach 50 g (1.8 oz) and 10 cm (3.9 in).

The longest beetle is the Hercules beetle *Dynastes hercules*, with a maximum overall length of at least 16.7 cm (6.6 in) including the very long pronotal horn. The smallest recorded beetle and the smallest free-living insect (as of 2015), is the featherwing beetle *Scydosella musawasensis* which may measure as little as 325 μm in length.

Hercules beetle, Dynastes hercules ecuatorianus, the longest of all beetles.

Titan beetle, *Titanus giganteus*, a tropical longhorn, is one of the largest and heaviest insects in the world.

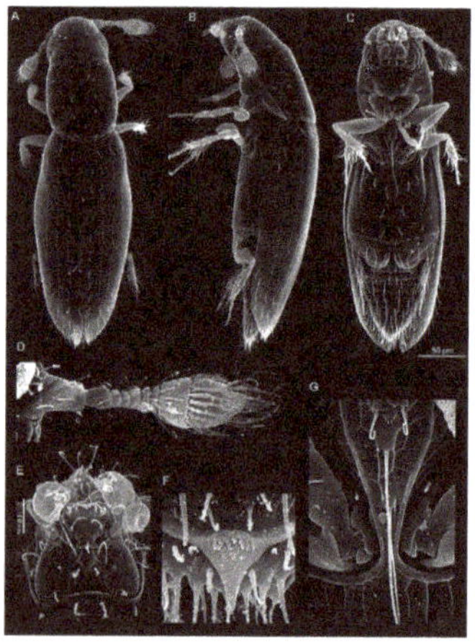

Scydosella musawasensis, the smallest known beetle: scale bar (right) is 50 µm.

Evolution

Late Paleozoic

The oldest known fossil insect that unequivocally resembles a Coleopteran is from the Lower Permian Period about 270 million years ago (mya), though these members of the family Tshekardocoleidae have 13-segmented antennae, elytra with more fully developed venation and more irregular longitudinal ribbing, and abdomen and ovipositor extending beyond the apex of the elytra. In the Permian–Triassic extinction event at the end of the Permian, some 30% of all insect species became extinct, so the fossil record of insects only includes beetles from the Lower Triassic 220 mya. Around this time, during the Late Triassic, fungus-feeding species such as Cupedidae appear in the fossil record. In the stages of the Upper Triassic, alga-feeding insects such as Triaplidae and Hydrophilidae begin to appear, alongside predatory water beetles. The first weevils, including the Obrienidae, appear alongside the first rove beetles (Staphylinidae), which closely resemble recent species. Some entomologists are sceptical that such early insects are so closely related to present-day species, arguing that this is extremely unlikely; for example, the structure of the metepisternum suggests that the Obrienidae could be Archostemata, not weevils at all, despite fossils with weevil-like snouts.

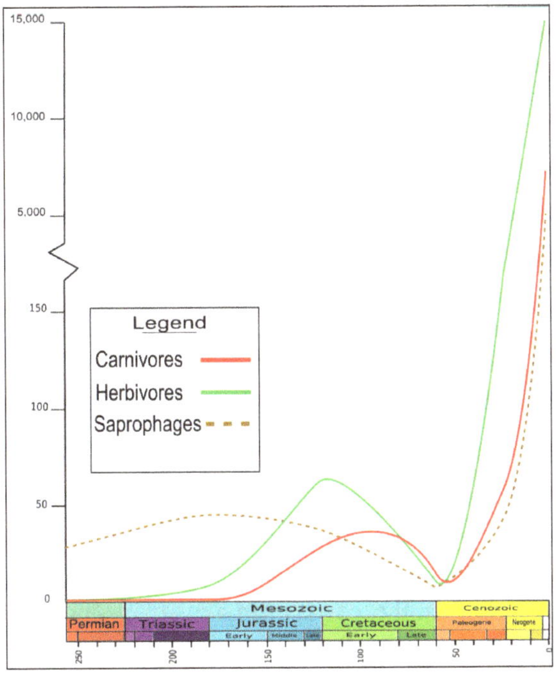

Beetle genera were mainly saprophages (detritivores) in the Permian and Triassic. During the Jurassic, herbivorous and then carnivorous genera became more common. In the Cenozoic, genera at all three trophic levels became far more numerous.

In 2009, a fossil beetle was described from the Pennsylvanian of Mazon Creek, Illinois, pushing the origin of the beetles to an earlier date, 318 to 299 mya. Fossils from this time have been found in Asia and Europe, for instance in the red slate fossil beds of Niedermoschel near Mainz, Germany. Further fossils have been found in Obora, Czech Republic and Tshekarda in the Ural mountains, Russia. However, there are only a few fossils from North America before the middle Permian, although both Asia and North America had been united to Euramerica. The first discoveries from North America made in the Wellington formation of Oklahoma were published in 2005 and 2008.

As a consequence of the Permian–Triassic extinction event, the fossil record of insects is scant, including beetles from the Lower Triassic. However, there are a few exceptions, such as in Eastern Europe. At the Babiy Kamen site in the Kuznetsk Basin, numerous beetle fossils were discovered, including entire specimens of the infraorders Archostemata (e.g. Ademosynidae, Schizocoleidae), Adephaga (e.g., Triaplidae, Trachypachidae) and Polyphaga (e.g. Hydrophilidae, Byrrhidae, Elateroidea). However, species from the families Cupedidae and Schizophoroidae are not present at this site, whereas they dominate at other fossil sites from the Lower Triassic such as Khey-Yaga, Russia, in the Korotaikha Basin.

Jurassic

During the Jurassic (210 to 145 mya), there was a dramatic increase in the diversity of beetle families, including the development and growth of carnivorous and herbivorous species. The Chrysomeloidea diversified around the same time, feeding on a wide array of plant hosts from cycads and conifers to angiosperms. Close to the Upper Jurassic, the Cupedidae decreased, but the diversity of the early plant-eating species increased. Most recent plant-eating beetles feed on flowering plants or angiosperms, whose success contributed to a doubling of plant-eating species during the Middle Jurassic. However, the increase of the number of beetle families during the Cretaceous does not correlate with the increase of the number of angiosperm species. Around the same time, numerous primitive weevils (e.g. Curculionoidea) and click beetles (e.g. Elateroidea) appeared. The first jewel beetles (e.g. Buprestidae) are present, but they remained rare until the Cretaceous. The first scarab beetles were not coprophagous but presumably fed on rotting wood with the help of fungus; they are an early example of a mutualistic relationship.

There are more than 150 important fossil sites from the Jurassic, the majority in Eastern Europe and North Asia. Outstanding sites include Solnhofen in Upper Bavaria, Germany, Karatau in South Kazakhstan, the Yixian formation in Liaoning, North China, as well as the Jiulongshan formation and further fossil sites in Mongolia. In North America there are only a few sites with fossil records of insects from the Jurassic, namely the shell limestone deposits in the Hartford basin, the Deerfield basin and the Newark basin.

Cretaceous

The Cretaceous saw the fragmenting of the southern landmass, with the opening of the southern Atlantic Ocean and the isolation of New Zealand, while South America, Antarctica, and Australia grew more distant. The diversity of Cupedidae and Archostemata decreased considerably. Predatory ground beetles (Carabidae) and rove beetles (Staphylinidae) began to distribute into different patterns; the Carabidae predominantly occurred in the warm regions, while the Staphylinidae and click beetles (Elateridae) preferred temperate climates. Likewise, predatory species of Cleroidea and Cucujoidea hunted their prey under the bark of trees together with the jewel beetles (Buprestidae). The diversity of jewel beetles increased rapidly, as they were the primary consumers of wood, while longhorn beetles (Cerambycidae) were rather rare: their diversity increased only towards the end of the Upper Cretaceous. The first coprophagous beetles are from the Upper Cretaceous and may have lived on the excrement of herbivorous dinosaurs. The first species where both larvae and adults are adapted to an aquatic lifestyle are found. Whirligig beetles (Gyrinidae) were moderately diverse, although other early beetles (e.g. Dytiscidae) were less, with the most widespread being the species of Coptoclavidae, which preyed on aquatic fly larvae.

Many fossil sites worldwide contain beetles from the Cretaceous. Most are in Europe and Asia and belong to the temperate climate zone during the Cretaceous. Lower

The elytra allow beetles to both fly and move through confined spaces, doing so by folding the delicate wings under the elytra while not flying, and folding their wings out just before takeoff. The unfolding and folding of the wings is operated by muscles attached to the wing base; as long as the tension on the radial and cubital veins remains, the wings remain straight. In some day-flying species (for example, Buprestidae, Scarabaeidae), flight does not include large amounts of lifting of the elytra, having the metathorac wings extended under the lateral elytra margins. The altitude reached by beetles in flight varies. One study investigating the flight altitude of the ladybird species *Coccinella septempunctata* and *Harmonia axyridis* using radar showed that, whilst the majority in flight over a single location were at 150–195 m above ground level, some reached altitudes of over 1100 m.

Many rove beetles have greatly reduced elytra, and while they are capable of flight, they most often move on the ground: their soft bodies and strong abdominal muscles make them flexible, easily able to wriggle into small cracks.

Aquatic beetles use several techniques for retaining air beneath the water's surface. Diving beetles (Dytiscidae) hold air between the abdomen and the elytra when diving. Hydrophilidae have hairs on their under surface that retain a layer of air against their bodies. Adult crawling water beetles use both their elytra and their hind coxae (the basal segment of the back legs) in air retention, while whirligig beetles simply carry an air bubble down with them whenever they dive.

Communication

Beetles have a variety of ways to communicate, including the use of pheromones. The mountain pine beetle emits a pheromone to attract other beetles to a tree. The mass of beetles are able to overcome the chemical defenses of the tree. After the tree's defenses have been exhausted, the beetles emit an anti-aggregation pheromone. This species can stridulate to communicate, but others may use sound to defend themselves when attacked.

Parental Care

A dung beetle rolling dung.

Parental care is found in a few families of beetle, perhaps for protection against adverse conditions and predators. The rove beetle *Bledius spectabilis* lives in salt marshes, so the eggs and larvae are endangered by the rising tide. The maternal beetle patrols the eggs and larvae, burrowing to keep them from flooding and asphyxiating, and protects them from the predatory carabid beetle *Dicheirotrichus gustavi* and from the parasitoidal wasp *Barycnemis blediator*, which kills some 15% of the larvae.

Burying beetles are attentive parents, and participate in cooperative care and feeding of their offspring. Both parents work to bury small animal carcass to serve as a food resource for their young and build a brood chamber around it. The parents prepare the carcass and protect it from competitors and from early decomposition. After their eggs hatch, the parents keep the larvae clean of fungus and bacteria and help the larvae feed by regurgitating food for them.

Some dung beetles provide parental care, collecting herbivore dung and laying eggs within that food supply, an instance of mass provisioning. Some species do not leave after this stage, but remain to safeguard their offspring.

Most species of beetles do not display parental care behaviors after the eggs have been laid.

Subsociality, where females guard their offspring, is well-documented in two families of Chrysomelidae, Cassidinae and Chrysomelinae.

Eusociality

Eusociality involves cooperative brood care (including brood care of offspring from other individuals), overlapping generations within a colony of adults, and a division of labour into reproductive and non-reproductive groups. Few organisms outside Hymenoptera exhibit this behavior; the only beetle to do so is the weevil *Austroplatypus incompertus*. This Australian species lives in horizontal networks of tunnels, in the heartwood of *Eucalyptus* trees. It is one of more than 300 species of wood-boring Ambrosia beetles which distribute the spores of ambrosia fungi. The fungi grow in the beetles' tunnels, providing food for the beetles and their larvae; female offspring remain in the tunnels and maintain the fungal growth, probably never reproducing. Cooperative brood care is also found in the bess beetles (Passalidae) where the larvae feed on the semi-digested faeces of the adults.

Feeding

Beetles are able to exploit a wide diversity of food sources available in their many habitats. Some are omnivores, eating both plants and animals. Other beetles are highly specialized in their diet. Many species of leaf beetles, longhorn beetles, and weevils are very host-specific, feeding on only a single species of plant. Ground beetles and rove beetles (Staphylinidae), among others, are primarily carnivorous and catch and consume many other arthropods and small prey, such as earthworms and snails. While most predatory beetles are generalists, a few species have more specific prey requirements or preferences.

Hycleus sp. (Meloidae) feeding on the petals of *Ipomoea carnea*.

Decaying organic matter is a primary diet for many species. This can range from dung, which is consumed by coprophagous species (such as certain scarab beetles in the Scarabaeidae), to dead animals, which are eaten by necrophagous species (such as the carrion beetles, Silphidae). Some beetles found in dung and carrion are in fact predatory. These include members of the Histeridae and Silphidae, preying on the larvae of coprophagous and necrophagous insects. Many beetles feed under bark, some feed on wood while others feed on fungi growing on wood or leaf-litter. Some beetles have special mycangia, structures for the transport of fungal spores.

Ecology

Anti-predator Adaptations

A camouflaged longhorn beetle, *Ecyrus dasycerus*.

Beetles, both adults and larvae, are the prey of many animal predators including mammals from bats to rodents, birds, lizards, amphibians, fishes, dragonflies, robberflies, reduviid bugs, ants, other beetles, and spiders. Beetles use a variety of anti-predator adaptations to defend themselves. These include camouflage and mimicry against predators that hunt by sight, toxicity, and defensive behaviour.

Camouflage

Camouflage is common and widespread among beetle families, especially those that feed on wood or vegetation, such as leaf beetles (Chrysomelidae, which are often green) and weevils. In some species, sculpturing or various coloured scales or hairs cause beetles such as the avocado weevil *Heilipus apiatus* to resemble bird dung or other inedible objects. Many beetles that live in sandy environments blend in with the coloration of that substrate.

Mimicry and Aposematism

Clytus arietis (Cerambycidae), a Batesian mimic of wasps.

Some longhorn beetles (Cerambycidae) are effective Batesian mimics of wasps. Beetles may combine coloration with behavioural mimicry, acting like the wasps they already closely resemble. Many other beetles, including ladybirds, blister beetles, and lycid beetles secrete distasteful or toxic substances to make them unpalatable or poisonous, and are often aposematic, where bright or contrasting coloration warn off predators; many beetles and other insects mimic these chemically protected species.

Blister beetles such as *Hycleus* have brilliant aposematic coloration, warning of their toxicity.

The bloody-nosed beetle, Timarcha tenebricosa, defending itself by releasing a droplet of noxious red liquid.

Chemical defense is important in some species, usually being advertised by bright aposematic colours. Some Tenebrionidae use their posture for releasing noxious chemicals to warn off predators. Chemical defences may serve purposes other than just protection from vertebrates, such as protection from a wide range of microbes. Some species sequester chemicals from the plants they feed on, incorporating them into their own defenses.

Other species have special glands to produce deterrent chemicals. The defensive glands of carabid ground beetles produce a variety of hydrocarbons, aldehydes, phenols, quinones, esters, and acids released from an opening at the end of the abdomen. African carabid beetles (for example, *Anthia* and *Thermophilum* – *Thermophilum* is sometimes included within *Anthia*) employ the same chemicals as ants: Formic acid. Bombardier beetles have well-developed pygidial glands that empty from the sides of the intersegment membranes between the seventh and eighth abdominal segments. The gland is made of two containing chambers, one for hydroquinones and hydrogen peroxide, the other holding hydrogen peroxide and catalase enzymes. These chemicals mix and result in an explosive ejection, reaching a temperature of around 100 °C (212 °F), with the breakdown of hydroquinone to hydrogen, oxygen, and quinone. The oxygen propels the noxious chemical spray as a jet that can be aimed accurately at predators.

Other Defences

Large ground-dwelling beetles such as Carabidae, the rhinoceros beetle and the longhorn beetles defend themselves using strong mandibles, or heavily sclerotised (armored) spines or horns to deter or fight off predators. Many species of weevil that feed out in the open on leaves of plants react to attack by employing a drop-off reflex. Some combine it with thanatosis, in which they close up their appendages and "play dead". The click beetles (Elateridae) can suddenly catapult themselves out of danger by releasing the energy stored by a click mechanism, which consists of a stout spine on the prosternum and a matching groove in the mesosternum. Some species startle an attacker by producing sounds through a process known as stridulation.

Parasitism

A few species of beetles are ectoparasitic on mammals. One such species, *Platypsyllus castoris,* parasitises beavers (*Castor* spp.). This beetle lives as a parasite both as a larva and as an adult, feeding on epidermal tissue and possibly on skin secretions and wound exudates. They are strikingly flattened dorsoventrally, no doubt as an adaptation for slipping between the beavers' hairs. They are wingless and eyeless, as are many other ectoparasites. Others are kleptoparasites of other invertebrates, such as the small hive beetle (*Aethina tumida*) that infests honey bee nests, while many species are parasitic inquilines or commensal in the nests of ants. A few groups of beetles are primary parasitoids of other insects, feeding off of, and eventually killing their hosts.

Pollination

An Israeli Copper Flower-Chafer (Protaetia cuprea ignicollis) pollinating a crown daisy (Glebionis coronaria).

Beetle-pollinated flowers are usually large, greenish or off-white in color, and heavily scented. Scents may be spicy, fruity, or similar to decaying organic material. Beetles were most likely the first insects to pollinate flowers. Most beetle-pollinated flowers are flattened or dish-shaped, with pollen easily accessible, although they may include traps to keep the beetle longer. The plants' ovaries are usually well protected from the biting mouthparts of their pollinators. The beetle families that habitually pollinate flowers are the Buprestidae, Cantharidae, Cerambycidae, Cleridae, Dermestidae, Lycidae, Melyridae, Mordellidae, Nitidulidae and Scarabaeidae. Beetles may be particularly important in some parts of the world such as semiarid areas of southern Africa and southern California and the montane grasslands of KwaZulu-Natal in South Africa.

Mutualism

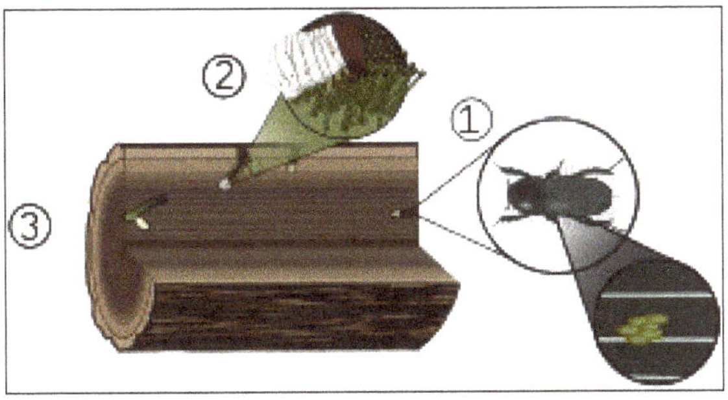

1: Adult ambrosia beetle burrows into wood and lays eggs, carrying fungal spores in its mycangia.
2: Larva feeds on fungus, which digests wood, removing toxins, to mutual benefit. 3: Larva pupates.

Mutualism is well known in a few beetles, such as the ambrosia beetle, which partners with fungi to digest the wood of dead trees. The beetles excavate tunnels in dead trees in which they cultivate fungal gardens, their sole source of nutrition. After landing on a suitable tree, an ambrosia beetle excavates a tunnel in which it releases spores of its fungal symbiont. The fungus penetrates the plant's xylem tissue, digests it, and concentrates the nutrients on and near the surface of the beetle gallery, so the weevils and the fungus both benefit. The beetles cannot eat the wood due to toxins, and uses its relationship with fungi to help overcome the defenses of its host tree in order to provide nutrition for their larvae. Chemically mediated by a bacterially produced polyunsaturated peroxide, this mutualistic relationship between the beetle and the fungus is coevolved.

Beetle found in Thar Desert.

Tolerance of Extreme Environments

About 90% of beetle species enter a period of adult diapause, a quiet phase with reduced metabolism to tide unfavourable environmental conditions. Adult diapause is the most common form of diapause in Coleoptera. To endure the period without food (often lasting many months) adults prepare by accumulating reserves of lipids, glycogen, proteins and other substances needed for resistance to future hazardous changes of environmental conditions. This diapause is induced by signals heralding the arrival of the unfavourable season; usually the cue is photoperiodic. Short (decreasing) day length serves as a signal of approaching winter and induces winter diapause (hibernation). A study of hibernation in the Arctic beetle *Pterostichus brevicorni* showed that the body fat levels of adults were highest in autumn with the alimentary canal filled with food, but empty by the end of January. This loss of body fat was a gradual process, occurring in combination with dehydration.

The fogstand beetle of the Namib Desert, Stenocara gracilipes is able to survive by collecting water from fog on its back.

All insects are poikilothermic, so the ability of a few beetles to live in extreme environments depends on their resilience to unusually high or low temperatures. The bark beetle *Pityogenes chalcographus* can survive −39 °C whilst overwintering beneath tree bark; the Alaskan beetle *Cucujus clavipes puniceus* is able to withstand −58 °C; its larvae may survive −100 °C. At these low temperatures, the formation of ice crystals in internal fluids is the biggest threat to survival to beetles, but this is prevented through the production of antifreeze proteins that stop water molecules from grouping together. The low temperatures experienced by *Cucujus clavipes* can be survived through their deliberate dehydration in conjunction with the antifreeze proteins. This concentrates the antifreezes several fold. The hemolymph of the mealworm beetle *Tenebrio molitor* contains several antifreeze proteins. The Alaskan beetle *Upis ceramboides* can survive −60 °C: its cryoprotectants are xylomannan, a molecule consisting of a sugar bound to a fatty acid, and the sugar-alcohol, threitol.

Conversely, desert dwelling beetles are adapted to tolerate high temperatures. For example, the Tenebrionid beetle *Onymacris rugatipennis* can withstand 50 °C. Tiger beetles in hot, sandy areas are often whitish (for example, *Habroscelimorpha dorsalis*), to reflect more heat than a darker colour would. These beetles also exhibits behavioural adaptions to tolerate the heat: they are able to stand erect on their tarsi to hold their bodies away from the hot ground, seek shade, and turn to face the sun so that only the front parts of their heads are directly exposed.

The fogstand beetle of the Namib Desert, *Stenocara gracilipes*, is able to collect water from fog, as its elytra have a textured surface combining hydrophilic (water-loving) bumps and waxy, hydrophobic troughs. The beetle faces the early morning breeze, holding up its abdomen; droplets condense on the elytra and run along ridges towards their mouthparts. Similar adaptations are found in several other Namib desert beetles such as *Onymacris unguicularis*.

Some terrestrial beetles that exploit shoreline and floodplain habitats have physiological adaptations for surviving floods. In the event of flooding, adult beetles may be mobile enough to move away from flooding, but larvae and pupa often cannot. Adults of *Cicindela togata* are unable to survive immersion in water, but larvae are able to survive a prolonged period, up to 6 days, of anoxia during floods. Anoxia tolerance in the larvae may have been sustained by switching to anaerobic metabolic pathways or by reducing metabolic rate. Anoxia tolerance in the adult Carabid beetle *Pelophilia borealis* was tested in laboratory conditions and it was found that they could survive a continuous period of up to 127 days in an atmosphere of 99.9% nitrogen at 0 °C.

Migration

Many beetle species undertake annual mass movements which are termed as migrations. These include the pollen beetle *Meligethes aeneus* and many species of coccinellids. These mass movements may also be opportunistic, in search of food, rather than seasonal. A 2008 study of an unusually large outbreak of Mountain Pine Beetle (*Dendroctonus ponderosae*) in British Columbia found that beetles were capable of flying 30–110 km per day in densities of up to 18, 600 beetles per hectare.

DIPTERA

True flies are insects of the order Diptera. Insects of this order use only a single pair of wings to fly, the hindwings having evolved into advanced mechanosensory organs known as halteres, which act as high-speed sensors of rotational movement and allow dipterans to perform advanced aerobatics. Diptera is a large order containing an estimated 1,000,000 species including horse-flies, crane flies, hoverflies and others, although only about 125,000 species have been described.

Flies have a mobile head, with a pair of large compound eyes, and mouthparts designed for piercing and sucking (mosquitoes, black flies and robber flies), or for lapping and sucking in the other groups. Their wing arrangement gives them great maneuverability in flight, and claws and pads on their feet enable them to cling to smooth surfaces. Flies undergo complete metamorphosis; the eggs are laid on the larval food-source and the larvae, which lack true limbs, develop in a protected environment, often inside their food source. The pupa is a tough capsule from which the adult emerges when ready to do so; flies mostly have short lives as adults.

Diptera is one of the major insect orders and of considerable ecological and human importance. Flies are important pollinators, second only to the bees and their Hymenopteran relatives. Flies may have been among the evolutionarily earliest pollina-tors responsible for early plant pollination. Fruit flies are used as model organisms in research, but less benignly, mosquitoes are vectors for malaria, dengue, West Nile fever, yellow fever, encephalitis, and other infectious diseases; and houseflies, com-mensal with humans all over the world, spread food-borne illnesses. Flies can be an-noyances especially in some parts of the world where they can occur in large numbers, buzzing and settling on the skin or eyes to bite or seek fluids. Larger flies such as tsetse flies and screwworms cause significant economic harm to cattle. Blowfly larvae, known as gentles, and other dipteran larvae, known more generally as maggots, are used as fishing bait and as food for carnivorous animals. They are also used in medicine in de-bridement to clean wounds.

Taxonomy and Phylogeny

Relationships to other Insects

Dipterans are endopterygotes, insects that undergo radical metamorphosis. They be-long to the Mecopterida, alongside the Mecoptera, Siphonaptera, Lepidoptera and Trichoptera. The possession of a single pair of wings distinguishes most true flies from other insects with "fly" in their names. However, some true flies such as Hippoboscidae (louse flies) have become secondarily wingless.

Relationships between Fly Subgroups and Families

Fossil brachyceran in Baltic amber. Lower Eocene, million years ago.

The first true dipterans known are from the Middle Triassic (around 240 million years ago), and they became widespread during the Middle and Late Triassic. Modern flow-ering plants did not appear until the Cretaceous (around 140 million years ago), so the original dipterans must have had a different source of nutrition other than nectar. Based on the attraction of many modern fly groups to shiny droplets, it has been sug-gested that they may have fed on honeydew produced by sap-sucking bugs which were abundant at the time, and dipteran mouthparts are well-adapted to softening and lapping up the crusted residues. The basal clades in the Diptera include the Deuterophlebiidae and the enigmatic Nymphomyiidae. Three episodes of evolutionary radiation are thought to have occurred based on the fossil record. Many new species of lower Diptera developed in the Triassic, about 220 million years ago. Many lower Brachycera appeared in the Jurassic, some 180 million years ago. A third radiation took place among the Schizophora at the start of the Paleogene, 66 million years ago.

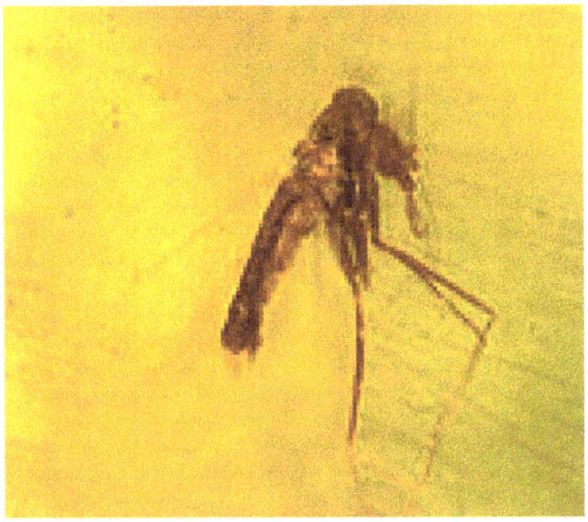

Fossil nematoceran in Dominican amber. Sandfly, Lutzomyia adiketis (Psychodidae), Early Miocene, million years ago.

The phylogenetic position of Diptera has been controversial. The monophyly of holometabolous insects has long been accepted, with the main orders being established as Lepidoptera, Coleoptera, Hymenoptera and Diptera, and it is the relationships between these groups which has caused difficulties. Diptera is widely thought to be a member of Mecopterida, along with Lepidoptera (butterflies and moths), Trichoptera (caddisflies), Siphonaptera (fleas), Mecoptera (scorpionflies) and possibly Strepsiptera (twisted-wing flies). Diptera has been grouped with Siphonaptera and Mecoptera in the Antliophora, but this has not been confirmed by molecular studies.

Diptera were traditionally broken down into two suborders, Nematocera and Brachycera, distinguished by the differences in antennae. The Nematocera are identified by their elongated bodies and many-segmented, often feathery antennae as represented by mosquitoes and crane flies. The Brachycera have rounder bodies and much shorter antennae. Subsequent studies have identified the Nematocera as being non-monophyletic with modern phylogenies placing the Brachycera within grades of groups formerly placed in the Nematocera. The construction of a phylogenetic tree has been the subject of ongoing research. The following cladogram is based on the FLYTREE project.

Diversity

Flies are often abundant and are found in almost all terrestrial habitats in the world apart from Antarctica. They include many familiar insects such as house flies, blow flies, mosquitoes, gnats, black flies, midges and fruit flies. More than 150,000 have been formally described and the actual species diversity is much greater, with the flies from many parts of the world yet to be studied intensively. The suborder Nematocera include generally small, slender insects with long antennae such as mosquitoes, gnats, midges and crane-flies, while the Brachycera includes broader, more robust flies with short antennae. Many nematoceran larvae are aquatic. There are estimated to be a total of about 19,000 species of Diptera in Europe, 22,000 in the Nearctic region, 20,000 in the Afrotropical region, 23,000 in the Oriental region and 19,000 in the Australasian region. While most species have restricted distributions, a few like the housefly (*Musca domestica*) are cosmopolitan. *Gauromydas heros* (Asiloidea), with a length of up to 7 cm (2.8 in), is generally considered to be the largest fly in the world, while the smallest is *Euryplatea nanaknihali*, which at 0.4 mm (0.016 in) is smaller than a grain of salt.

Gauromydas heros is the largest fly in the world.

Brachycera are ecologically very diverse, with many being predatory at the larval stage and some being parasitic. Animals parasitised include molluscs, woodlice, millipedes, insects, mammals, and amphibians. Flies are the second largest group of pollinators after the Hymenoptera (bees, wasps and relatives). In wet and colder environments flies are significantly more important as pollinators. Compared to bees, they need less food as they do not need to provision their young. Many flowers that bear low nectar and those that have evolved trap pollination depend on flies. It is thought that some of the earliest pollinators of plants may have been flies.

The greatest diversity of gall forming insects are found among the flies, principally in the family Cecidomyiidae (gall midges). Many flies (most importantly in the family Agromyzidae) lay their eggs in the mesophyll tissue of leaves with larvae feeding between the surfaces forming blisters and mines. Some families are mycophagous or fungus feeding. These include the cave dwelling Mycetophilidae (fungus gnats) whose larvae are the only diptera with bioluminescence. The Sciaridae are also fungus feeders. Some plants are pollinated by fungus feeding flies that visit fungus infected male flowers.

The larvae of *Megaselia scalaris* (Phoridae) are almost omnivorous and consume such substances as paint and shoe polish. The *Exorista mella (Walker)* fly are considered generalists and parasitoids of a variety of hosts. The larvae of the shore flies (Ephydridae) and some Chironomidae survive in extreme environments including glaciers (*Diamesa* sp., Chironomidae), hot springs, geysers, saline pools, sulphur pools, septic tanks and even crude oil (*Helaeomyia petrolei*). Adult hoverflies (Syrphidae) are well known for their mimicry and the larvae adopt diverse lifestyles including being inquiline scavengers inside the nests of social insects. Some brachycerans are agricultural pests, some bite animals and humans and suck their blood, and some transmit diseases.

Ecology

As ubiquitous insects, dipterans play an important role at various trophic levels both as consumers and as prey. In some groups the larvae complete their development without feeding, and in others the adults do not feed. The larvae can be herbivores, scavengers, decomposers, predators or parasites, with the consumption of decaying organic matter being one of the most prevalent feeding behaviours. The fruit or detritus is consumed along with the associated micro-organisms, a sieve-like filter in the pharynx being used to concentrate the particles, while flesh-eating larvae have mouth-hooks to help shred their food. The larvae of some groups feed on or in the living tissues of plants and fungi, and some of these are serious pests of agricultural crops. Some aquatic larvae consume the films of algae that form underwater on rocks and plants. Many of the parasitoid larvae grow inside and eventually kill other arthropods, while parasitic larvae may attack vertebrate hosts.

Whereas many dipteran larvae are aquatic or live in enclosed terrestrial locations, the majority of adults live above ground and are capable of flight. Predominantly they feed on nectar or plant or animal exudates, such as honeydew, for which their lapping mouthparts are adapted. The flies that feed on vertebrate blood have sharp stylets that pierce the skin, the insects inserting anticoagulant saliva and absorbing the blood that flows; in this process, certain diseases can be transmitted. The bot flies (Oestridae) have evolved to parasitize mammals. Many species complete their life cycle inside the bodies of their hosts. In many dipteran groups, swarming is a feature of adult life, with clouds of insects gathering in certain locations; these insects are mostly males, and the swarm may serve the purpose of making their location more visible to females.

Anti-predator Adaptations

Flies are eaten by other animals at all stages of their development. The eggs and larvae are parasitised by other insects and are eaten by many creatures, some of which specialise in feeding on flies but most of which consume them as part of a mixed diet. Birds, bats, frogs, lizards, dragonflies and spiders are among the predators of flies. Many flies have evolved mimetic resemblances that aid their protection. Batesian mimicry is widespread with many hoverflies resembling bees and wasps, ants and some species of tephritid fruit fly resembling spiders. Some species of hoverfly are myrmecophilous, their young live and grow within the nests of ants. They are protected from the ants by imitating chemical odours given by ant colony members. Bombyliid bee flies such as *Bombylius major* are short-bodied, round, furry, and distinctly bee-like as they visit flowers for nectar, and are likely also Batesian mimics of bees.

The large bee-fly, Bombylius major, is a Batesian mimic of bees.

In contrast, *Drosophila subobscura,* a species of fly in the genus *Drosophila,* lacks a category of hemocytes that are present in other studied species of *Drosophila,* leading to an inability to defend against parasitic attacks, a form of innate immunodeficiency.

Uses

Drosophila melanogaster fruit fly larvae being bred in tubes in a genetics laboratory.

Drosophila melanogaster, a fruit fly, has long been used as a model organism in research because of the ease with which it can be bred and reared in the laboratory, its small genome, and the fact that many of its genes have counterparts in higher eukaryotes. A large number of genetic studies have been undertaken based on this species; these have had a profound impact on the study of gene expression, gene regulatory mechanisms and mutation. Other studies have investigated physiology, microbial pathogenesis and development among other research topics. The studies on dipteran relationships by Willi Hennig helped in the development of cladistics, techniques that he applied to morphological characters but now adapted for use with molecular sequences in phylogenetics.

Blowflies feeding on the fresh corpse of a porcupine, *Hystrix africaeaustralis*.

Maggots found on corpses are useful to forensic entomologists. Maggot species can be identified by their anatomical features and by matching their DNA. Maggots of different species of flies visit corpses and carcases at fairly well-defined times after the death of the victim, and so do their predators, such as beetles in the family Histeridae. Thus, the presence or absence of particular species provides evidence for the time since death, and sometimes other details such as the place of death, when species are confined to particular habitats such as woodland.

Maggots used as animal feed at London Zoo.

Some species of maggots such as blowfly larvae (gentles) and bluebottle larvae (casters) are bred commercially; they are sold as bait in angling, and as food for carnivorous animals (kept as pets, in zoos, or for research) such as some mammals, fishes, reptiles, and birds. It has been suggested that fly larvae could be used at a large scale as food for farmed chickens, pigs, and fish. However, consumers are opposed to the inclusion of insects in their food, and the use of insects in animal feed remains illegal in areas such as the European Union.

Fly larvae can be used as a biomedical tool for wound care and treatment. Maggot debridement therapy (MDT) is the use of blow fly larvae to remove the dead tissue from wounds, most commonly being amputations. Historically, this has been used for centuries, both intentional and unintentional, on battlefields and in early hospital settings. Removing the dead tissue promotes cell growth and healthy wound healing. The larvae also have biochemical properties such as antibacterial activity found in their secretions as they feed. These medicinal maggots are a safe and effective treatment for chronic wounds.

Casu marzu is a traditional Sardinian sheep milk cheese that contains larvae of the cheese fly, *Piophila casei.*

The Sardinian cheese casu marzu is exposed to flies known as cheese skippers such as *Piophila casei*, members of the family Piophilidae. The digestive activities of the fly larvae soften the cheese and modify the aroma as part of the process of maturation. At one time European Union authorities banned sale of the cheese and it was becoming hard to find, but the ban has been lifted on the grounds that the cheese is a traditional local product made by traditional methods.

HYMENOPTERA

Hymenoptera is a large order of insects, comprising the sawflies, wasps, bees, and ants. Over 150,000 living species of Hymenoptera have been described, in addition to over 2,000 extinct ones.

Females typically have a special ovipositor for inserting eggs into hosts or places that are otherwise inaccessible. The ovipositor is often modified into a stinger. The young develop through holometabolism (complete metamorphosis)–that is, they have a worm-like larval stage and an inactive pupal stage before they mature.

Evolution

The cladogram of external relationships, based on a 2008 DNA and protein analysis, shows the order as a clade, most closely related to endopterygote orders including the Diptera (true flies) and Lepidoptera (butterflies and moths).

Hymenoptera originated in the Triassic, with the oldest fossils belonging to the family Xyelidae. Social hymenopterans appeared during the Cretaceous. The evolution of this group has been intensively studied by Alex Rasnitsyn, Michael S. Engel, and others.

This clade has been studied by examining the mitochondrial DNA. Although this study was unable to resolve all the ambiguities in this clade, some relationships could be established. The Aculeata, Ichneumonomorpha, and Proctotrupomorpha were monophyletic. The Megalyroidea and Trigonalyoidea are sister clades as are the Chalcidoidea+Diaprioidea. The Cynipoidea was generally recovered as the sister group to Chalcidoidea and Diaprioidea which are each other's closest relations.

Diet

Different species of Hymenoptera show a wide range of feeding habits. The most primitive forms are typically herbivorous, feeding on leaves or pine needles. Stinging wasps are predators, and will provision their larvae with immobilised prey, while bees feed on nectar and pollen.

A huge number of species are parasitoids as larvae. The adults inject the eggs into a host, which they begin to consume after hatching. For example, the eggs of the endangered *Papilio homerus* are parasitized at a rate of 77%, mainly by Hymenoptera species. Some species are even hyperparasitoid, with the host itself being another parasitoid insect. Habits intermediate between those of the herbivorous and parasitoid forms are shown in some hymenopterans, which inhabit the galls or nests of other insects, stealing their food, and eventually killing and eating the occupant.

Classification

Symphyta, without a waist: The sawfly *Arge pagana*.

Apocrita, with narrow waists: The wasp *Vespula germanica*.

The Hymenoptera are divided into two groups; the Symphyta which have no waist, and the Apocrita which have a narrow waist.

Symphyta

The suborder Symphyta includes the sawflies, horntails, and parasitic wood wasps. The group may be paraphyletic, as it has been suggested that the family Orussidae may be the group from which the Apocrita arose. They have an unconstricted junction between the thorax and abdomen. The larvae are herbivorous, free-living, and eruciform, with three pairs of true legs, prolegs (on every segment, unlike Lepidoptera) and ocelli. The prolegs do not have crochet hooks at the ends unlike the larvae of the Lepidoptera.

Apocrita

The wasps, bees, and ants together make up the suborder (and clade) Apocrita, characterized by a constriction between the first and second abdominal segments called a wasp-waist (petiole), also involving the fusion of the first abdominal segment to the thorax. Also, the larvae of all Apocrita lack legs, prolegs, or ocelli. The hindgut of the larvae also remains closed during development, with feces being stored inside the body, with the exception of some bee larvae where the larval anus has reappeared through developmental reversion. In general, the anus only opens at the completion of larval growth.

LEPIDOPTERA

Lepidoptera is an order of insects that includes butterflies and moths (both are called lepidopterans). About 180,000 species of the Lepidoptera are described, in 126 families and 46 superfamilies, 10 per cent of the total described species of living organisms. It is one of the most widespread and widely recognizable insect orders in the world. The Lepidoptera show many variations of the basic body structure that have evolved to gain advantages in lifestyle and distribution. Recent estimates suggest the order may have more species than earlier thought, and is among the four most speciose orders, along with the Hymenoptera, Diptera, and Coleoptera.

Lepidopteran species are characterized by more than three derived features. The most apparent is the presence of scales that cover the bodies, wings, and a proboscis. The scales are modified, flattened "hairs", and give butterflies and moths their wide variety of colors and patterns. Almost all species have some form of membranous wings, except for a few that have reduced wings or are wingless. Mating and the laying of eggs are carried out by adults, normally near or on host plants for the larvae. Like most other insects, butterflies and moths are holometabolous, meaning they undergo complete metamorphosis. The larvae are commonly called caterpillars, and are completely different from their adult moth or butterfly forms, having a cylindrical body with a well-developed head, mandible mouth parts, three pairs of thoracic legs and from none up to five pairs of prolegs. As they grow, these larvae change in appearance, going through a series of stages called instars. Once fully matured, the larva develops into a pupa. A few butterflies and many moth species spin a silk case or cocoon prior to pupating, while others do not, instead going underground. A butterfly pupa, called a chrysalis, has a hard skin, usually with no cocoon. Once the pupa has completed its metamorphosis, a sexually mature adult emerges.

The Lepidoptera have, over millions of years, evolved a wide range of wing patterns and coloration ranging from drab moths akin to the related order Trichoptera, to the brightly colored and complex-patterned butterflies. Accordingly, this is the most recognized and popular of insect orders with many people involved in the observation, study, collection, rearing of, and commerce in these insects. A person who collects or studies this order is referred to as a lepidopterist.

Butterflies and moths play an important role in the natural ecosystem as pollinators and as food in the food chain; conversely, their larvae are considered very problematic to vegetation in agriculture, as their main source of food is often live plant matter. In many species, the female may produce from 200 to 600 eggs, while in others, the number may approach 30,000 eggs in one day. The caterpillars hatching from these eggs can cause damage to large quantities of crops. Many moth and butterfly species are of economic interest by virtue of their role as pollinators, the silk they produce, or as pest species.

Distribution and Diversity

The Lepidoptera are among the most successful groups of insects. They are found on all continents, except Antarctica, and inhabit all terrestrial habitats ranging from desert to rainforest, from lowland grasslands to mountain plateaus, but almost always associated with higher plants, especially angiosperms (flowering plants). Among the most northern dwelling species of butterflies and moths is the Arctic Apollo (*Parnassius arcticus*), which is found in the Arctic Circle in northeastern Yakutia, at an altitude of 1500 m above sea level. In the Himalayas, various Apollo species such as *Parnassius epaphus* have been recorded to occur up to an altitude of 6,000 m above sea level.

Some lepidopteran species exhibit symbiotic, phoretic, or parasitic lifestyles, inhabiting the bodies of organisms rather than the environment. Coprophagous pyralid moth species, called sloth moths, such as *Bradipodicola hahneli* and *Cryptoses choloepi*, are unusual in that they are exclusively found inhabiting the fur of sloths, mammals found in Central and South America. Two species of *Tinea* moths have been recorded as feeding on horny tissue and have been bred from the horns of cattle. The larva of *Zenodochium coccivorella* is an internal parasite of the coccid *Kermes* species. Many species have been recorded as breeding in natural materials or refuse such as owl pellets, bat caves, honeycombs or diseased fruit.

As of 2007, there was roughly 174,250 lepidopteran species described, with butterflies and skippers estimated to comprise around 17,950, and moths making up the rest. The vast majority of Lepidoptera are to be found in the tropics, but substantial diversity exists on most continents. North America has over 700 species of butterflies and over 11,000 species of moths, while about 400 species of butterflies and 14,000 species of moths are reported from Australia. The diversity of Lepidoptera in each faunal region has been estimated by John Heppner in 1991 based partly on actual counts from the literature, partly on the card indices in the Natural History Museum (London) and the National Museum of Natural History (Washington), and partly on estimates:

Table: Diversity of Lepidoptera in each faunal region.

	Palearctic	Nearctic	Neotropic	Afrotropic	Indo-Australian (comprising Indomalayan and Australian regions)
Estimated number of species	22,465	11,532	44,791	20,491	47,287

Polymorphism

Sexually dimorphic bagworm moths (Thyridopteryx ephemeraeformis) mating: The female is flightless.

Polymorphism is the appearance of forms or "morphs", which differ in color and number of attributes within a single species. In Lepidoptera, polymorphism can be seen not only between individuals in a population, but also between the sexes as sexual dimorphism, between geographically separated populations in geographical polymorphism, and between generations flying at different seasons of the year (seasonal polymorphism or polyphenism). In some species, the polymorphism is limited to one sex, typically the female. This often includes the phenomenon of mimicry when mimetic morphs fly alongside nonmimetic morphs in a population of a particular species. Polymorphism occurs both at specific level with heritable variation in the overall morphological adaptations of individuals, as well as in certain specific morphological or physiological traits within a species.

The *Heliconius* butterflies from the tropics of the Western Hemisphere are the classical model for Müllerian mimicry.

Environmental polymorphism, in which traits are not inherited, is often termed as polyphenism, which in Lepidoptera is commonly seen in the form of seasonal morphs, especially in the butterfly families of Nymphalidae and Pieridae. An Old World pierid butterfly, the common grass yellow (*Eurema hecabe*) has a darker summer adult morph, triggered by a long day exceeding 13 hours in duration, while the shorter diurnal period of 12 hours or less induces a paler morph in the postmonsoon period. Polyphenism also occurs in caterpillars, an example being the peppered moth, *Biston betularia*.

Geographical isolation causes a divergence of a species into different morphs. A good example is the Indian white admiral *Limenitis procris*, which has five forms, each geographically separated from the other by large mountain ranges. An even more dramatic showcase of geographical polymorphism is the Apollo butterfly (*Parnassius apollo*). Because the Apollos live in small local populations, thus having no contact with each other, coupled with their strong stenotopic nature and weak migration ability, interbreeding between populations of one species practically does not occur; by this, they form over 600 different morphs, with the size of spots on the wings of which varies greatly.

Seasonal diphenism in the common grass yellow, *Eurema hecabe:* Dry-season form.

Wet-season form.

Sexual dimorphism is the occurrence of differences between males and females in a species. In Lepidoptera, it is widespread and almost completely set by genetic determination. Sexual dimorphism is present in all families of the Papilionoidea and more prominent in the Lycaenidae, Pieridae, and certain taxa of the Nymphalidae. Apart from color variation, which may differ from slight to completely different color-pattern combinations, secondary sexual characteristics may also be present. Different genotypes maintained by natural selection may also be expressed at the same time. Polymorphic and/or mimetic females occur in the case of some taxa in the Papilionidae primarily to obtain a level of protection not available to the male of their species. The most distinct case of sexual dimorphism is that of adult females of many Psychidae species which have only vestigial wings, legs, and mouthparts as compared to the adult males that are strong fliers with well-developed wings and feathery antennae.

Behavior

Flight

Flight is an important aspect of the lives of butterflies and moths, and is used for evading predators, searching for food, and finding mates in a timely manner, as lepidopteran species do not live long after eclosion. It is the main form of locomotion in most species. In Lepidoptera, the forewings and hindwings are mechanically coupled and flap in synchrony. Flight is anteromotoric, or being driven primarily by action of the forewings. Although lepidopteran species reportedly can still fly when their hindwings are cut off, it reduces their linear flight and turning capabilities.

Lepidopteran species have to be warm, about 77 to 79 °F (25 to 26 °C), to fly. They depend on their body temperature being sufficiently high and since they cannot regulate it themselves, this is dependent on their environment. Butterflies living in cooler climates may use their wings to warm their bodies. They will bask in the sun, spreading out their wings so that they get maximum exposure to the sunlight. In hotter climates butterflies can easily overheat, so they are usually active only during the cooler parts of the day, early morning, late afternoon or early evening. During the heat of the day, they rest in the shade. Some larger thick-bodied moths (e.g. Sphingidae) can generate their own heat to a limited degree by vibrating their wings. The heat generated by the flight muscles warms the thorax while the temperature of the abdomen is unimportant for flight. To avoid overheating, some moths rely on hairy scales, internal air sacs, and other structures to separate the thorax and abdomen and keep the abdomen cooler.

Some species of butterflies can reach fast speeds, such as the southern dart, which can go as fast as 48.4 km/h. Sphingids are some of the fastest flying insects, some are capable of flying at over 50 km/h (30 mi/h), having a wingspan of 35–150 mm. In some species, sometimes a gliding component to their flight exists. Flight occurs either as hovering, or as forward or backward motion. In butterfly and moth species, such as hawk moths, hovering is important as they need to maintain certain stability over flowers when feeding on the nectar.

Navigation

Timelapse of flying moths, attracted to the floodlights.

Navigation is important to Lepidoptera species, especially for those that migrate. Butterflies, which have more species that migrate, have been shown to navigate using time-compensated sun compasses. They can see polarized light, so can orient even in cloudy conditions. The polarized light in the region close to the ultraviolet spectrum is suggested to be particularly important. Most migratory butterflies are those that live in semiarid areas where breeding seasons are short. The life histories of their host plants also influence the strategies of the butterflies. Other theories include the use of landscapes. Lepidoptera may use coastal lines, mountains, and even roads to orient themselves. Above sea, the flight direction is much more accurate if the coast is still visible.

Many studies have also shown that moths navigate. One study showed that many moths may use the Earth's magnetic field to navigate, as a study of the moth heart and dart suggests. Another study, of the migratory behavior of the silver Y, showed, even at high altitudes, the species can correct its course with changing winds, and prefers flying with favourable winds, suggesting a great sense of direction. *Aphrissa statira* in Panama loses its navigational capacity when exposed to a magnetic field, suggesting it uses the Earth's magnetic field.

Moths exhibit a tendency to circle artificial lights repeatedly. This suggests they use a technique of celestial navigation called transverse orientation. By maintaining a constant angular relationship to a bright celestial light, such as the Moon, they can fly in a straight line. Celestial objects are so far away, even after traveling great distances, the change in angle between the moth and the light source is negligible; further, the moon will always be in the upper part of the visual field or on the horizon. When a moth encounters a much closer artificial light and uses it for navigation, the angle changes noticeably after only a short distance, in addition to being often below the horizon. The moth instinctively attempts to correct by turning toward the light, causing airborne moths to come plummeting downwards, and at close range, which results in a spiral flight path that gets closer and closer to the light source. Other explanations have been suggested, such as the idea that moths may be impaired with a visual distortion called a Mach band by Henry Hsiao in 1972. He stated that they fly towards the darkest part of the sky in pursuit of safety, thus are inclined to circle ambient objects in the Mach band region.

Migration

Monarch butterflies, seen in a cluster in Santa Cruz, California, where the western population migrates for the winter.

Lepidopteran migration is typically seasonal, as the insects moving to escape dry seasons or other disadvantageous conditions. Most lepidopterans that migrate are butterflies, and the distance travelled varies. Some butterflies that migrate include the mourning cloak, painted lady, American lady, red admiral, and the common buckeye. A notable species of moth that migrates long distances is the bogong moth. The most well-known migrations are those of the eastern population of the monarch butterfly from Mexico to northern United States and southern Canada, a distance of about 4,000–4,800 km (2,500–3,000 mi). Other well-known migratory species include the painted lady and several of the danaine butterflies. Spectacular and large-scale migrations associated with the monsoons are seen in peninsular India. Migrations have been studied in more recent times using wing tags and stable hydrogen isotopes.

Moths also undertake migrations, an example being the uraniids. *Urania fulgens* undergoes population explosions and massive migrations that may be not surpassed by any other insect in the Neotropics. In Costa Rica and Panama, the first population movements may begin in July and early August and depending on the year, may be very massive, continuing unabated for as long as five months.

Communication

Group of Melitaea athalia near Warka.

Pheromones are commonly involved in mating rituals among species, especially moths, but they are also an important aspect of other forms of communication. Usually, the pheromones are produced by either the male or the female and detected by members of the opposite sex with their antennae. In many species, a gland between the eighth and ninth segments under the abdomen in the female produces the pheromones. Communication can also occur through stridulation, or producing sounds by rubbing various parts of the body together.

Moths are known to engage in acoustic forms of communication, most often as courtship, attracting mates using sound or vibration. Like most other insects, moths pick up these sounds using tympanic membranes in their abdomens. An example is that of the polka-dot wasp moth (*Syntomeida epilais*), which produces sounds with a frequency above that normally detectable by humans (about 20 kHz). These sounds also function as tactile communication, or communication through touch, as they stridulate, or vibrate a substrate like leaves and stems.

Most moths lack bright colors, as many species use coloration as camouflage, but butterflies engage in visual communication. Female cabbage butterflies, for example, use ultraviolet light to communicate, with scales colored in this range on the dorsal wing surface. When they fly, each down stroke of the wing creates a brief flash of ultraviolet light which the males apparently recognize as the flight signature of a potential mate. These flashes from the wings may attract several males that engage in aerial courtship displays.

Ecology

Moths and butterflies are important in the natural ecosystem. They are integral participants in the food chain; having co-evolved with flowering plants and predators, lepidopteran species have formed a network of trophic relationships between autotrophs and heterotrophs, which are included in the stages of Lepidoptera larvae, pupae, and adults. Larvae and pupae are links in the diets of birds and parasitic entomophagous insects. The adults are included in food webs in a much broader range of consumers (including birds, small mammals, reptiles, etc).

Defense and Predation

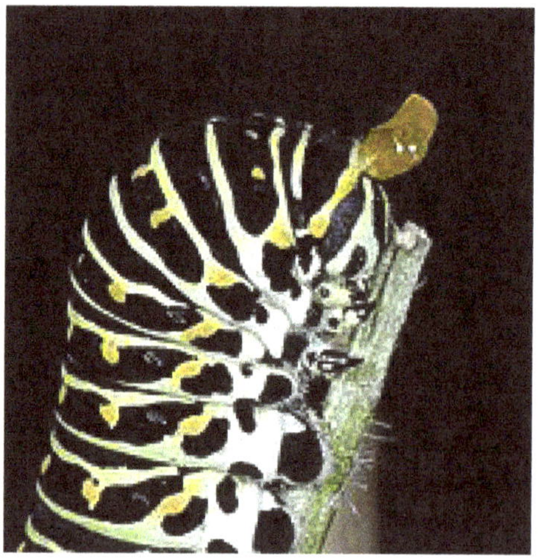

Papilio machaon caterpillar showing the osmeterium, which emits unpleasant smells to ward off predators.

Lepidopteran species are soft bodied, fragile, and almost defenseless, while the immature stages move slowly or are immobile, hence all stages are exposed to predation. Adult butterflies and moths are preyed upon by birds, bats, lizards, amphibians, dragonflies, and spiders. Caterpillars and pupae fall prey not only to birds, but also to invertebrate predators and small mammals, as well as fungi and bacteria. Parasitoid and parasitic wasps and flies may lay eggs in the caterpillar, which eventually kill it as they hatch inside its body and eat its tissues. Insect-eating birds are probably the largest predators. Lepidoptera, especially the immature stages, are an ecologically important food to many insectivorous birds, such as the great tit in Europe.

An "evolutionary arms race" can be seen between predator and prey species. The Lepidoptera have developed a number of strategies for defense and protection, including evolution of morphological characters and changes in ecological lifestyles and behaviors. These include aposematism, mimicry, camouflage, and development of threat patterns and displays. Only a few birds, such as the nightjars, hunt nocturnal lepidopterans. Their main predators are bats. Again, an "evolutionary race" exists, which has led to numerous evolutionary adaptations of moths to escape from their main predators, such as the ability to hear ultrasonic sounds, or even to emit sounds in some cases. Lepidopteran eggs are also preyed upon. Some caterpillars, such as the zebra swallowtail butterfly larvae, are cannibalistic.

Some species of Lepidoptera are poisonous to predators, such as the monarch butterfly in the Americas, *Atrophaneura* species (roses, windmills, etc.) in Asia, as well as *Papilio antimachus*, and the birdwings, the largest butterflies in Africa and Asia, respectively. They obtain their toxicity by sequestering the chemicals from the plants they eat into their own tissues. Some Lepidoptera manufacture their own toxins. Predators that eat poisonous butterflies and moths may become sick and vomit violently, learning not to eat those species. A predator which has previously eaten a poisonous lepidopteran may avoid other species with similar markings in the future, thus saving many other species, as well. Toxic butterflies and larvae tend to develop bright colors and striking patterns as an indicator to predators about their toxicity. This phenomenon is known as aposematism. Some caterpillars, especially members of Papilionidae, contain an osmeterium, a Y-shaped protrusible gland found in the prothoracic segment of the larvae. When threatened, the caterpillar emits unpleasant smells from the organ to ward off the predators.

Camouflage is also an important defense strategy, which involves the use of coloration or shape to blend into the surrounding environment. Some lepidopteran species blend with their surroundings, making them difficult to spot by predators. Caterpillars can exhibit shades of green that match its host plant. Others look like inedible objects, such as twigs or leaves. For instance, the mourning cloak fades into the backdrop of trees when it folds its wings back. The larvae of some species, such as the common Mormon (*Papilio polytes*) and the western tiger swallowtail look like bird droppings. For example, adult Sesiidae species (also known as clearwing moths) have a general appearance sufficiently similar to a wasp or hornet to make it likely the moths gain a reduction in predation by Batesian mimicry. Eyespots are a type of automimicry used by some butterflies and moths. In butterflies, the spots are composed of concentric rings of scales in different colors. The proposed role of the eyespots is to deflect attention of predators.

Their resemblance to eyes provokes the predator's instinct to attack these wing patterns.

Batesian and Müllerian mimicry complexes are commonly found in Lepidoptera. Genetic polymorphism and natural selection give rise to otherwise edible species (the mimic) gaining a survival advantage by resembling inedible species (the model). Such a mimicry complex is referred to as Batesian and is most commonly known in the example between the limenitidine viceroy butterfly in relation to the inedible danaine monarch. The viceroy is, in fact, more toxic than the monarch and this resemblance should be considered as a case of Müllerian mimicry. In Müllerian mimicry, inedible species, usually within a taxonomic order, find it advantageous to resemble each other so as to reduce the sampling rate by predators that need to learn about the insects' inedibility. Taxa from the toxic genus *Heliconius* form one of the most well-known Müllerian complexes. The adults of the various species now resemble each other so well, the species cannot be distinguished without close morphological observation and, in some cases, dissection or genetic analysis.

Moths evidently are able to hear the range emitted by bats, which in effect causes flying moths to make evasive maneuvers because bats are a main predator of moths. Ultrasonic frequencies trigger a reflex action in the noctuid moth that cause it to drop a few inches in its flight to evade attack. Tiger moths in a defense emit clicks within the same range of the bats, which interfere with the bats and foil their attempts to echolocate it.

Pollination

A day-flying hummingbird hawk-moth drinking nectar from a species of *Dianthus*.

Most species of Lepidoptera engage in some form of entomophily (more specifically psychophily and phalaenophily for butterflies and moths, respectively), or the pollination of flowers. Most adult butterflies and moths feed on the nectar inside flowers, using their probosces to reach the nectar hidden at the base of the petals. In the process, the adults brush against the flowers' stamens, on which the reproductive pollen is made and stored. The pollen is transferred on appendages on the adults, which fly to the next flower to feed and unwittingly deposit the pollen on the stigma of the next flower, where the pollen germinates and fertilizes the seeds.

Flowers pollinated by butterflies tend to be large and flamboyant, pink or lavender in color, frequently having a landing area, and usually scented, as butterflies are typically day-flying. Since butterflies do not digest pollen (except for heliconid species,) more nectar is offered than pollen. The flowers have simple nectar guides, with the nectaries usually hidden in narrow tubes or spurs, reached by the long "tongue" of the butterflies. Butterflies such as *Thymelicus flavus* have been observed to engage in flower constancy, which means they are more likely to transfer pollen to other conspecific plants. This can be beneficial for the plants being pollinated, as flower constancy prevents the loss of pollen during different flights and the pollinators from clogging stigmas with pollen of other flower species.

Among the more important moth pollinator groups are the hawk moths of the family Sphingidae. Their behavior is similar to hummingbirds, i.e., using rapid wing beats to hover in front of flowers. Most hawk moths are nocturnal or crepuscular, so moth-pollinated flowers (e.g., *Silene latifolia*) tend to be white, night-opening, large, and showy with tubular corollae and a strong, sweet scent produced in the evening, night, or early morning. A lot of nectar is produced to fuel the high metabolic rates needed to power their flight. Other moths (e.g., noctuids, geometrids, pyralids) fly slowly and settle on the flower. They do not require as much nectar as the fast-flying hawk moths, and the flowers tend to be small (though they may be aggregated in heads).

Mutualism

Tobacco hornworm caterpillar (*Manduca sexta*) parasitized by Braconidae wasp larvae.

Mutualism is a form of biological interaction wherein each individual involved benefits in some way. An example of a mutualistic relationship would be that shared by yucca moths (Tegeculidae) and their host, yucca flowers (Liliaceae). Female yucca moths enter the host flowers, collect the pollen into a ball using specialized maxillary palps, then move to the apex of the pistil, where pollen is deposited on the stigma, and lay eggs into the base of the pistil where seeds will develop. The larvae develop in the fruit pod and feed on a portion of the seeds. Thus, both insect and plant benefit, forming a highly mutualistic relationship. Another form of mutualism occurs between some larvae of butterflies and certain species of ants (e. g. Lycaenidae). The larvae communicate with the ants using vibrations transmitted through a substrate, such as the wood of a tree or stems, as well as using chemical signals. The ants provide some degree of protection to these larvae and they in turn gather honeydew secretions.

Parasitism

Parasitoid larvae exits from the fox moth caterpillar.

Only 42 species of parasitoid lepidopterans are known (1 Pyralidae; 40 Epipyropidae). The larvae of the greater and lesser wax moths feed on the honeycomb inside bee nests and may become pests; they are also found in bumblebee and wasp nests, albeit to a lesser extent. In northern Europe, the wax moth is regarded as the most serious parasitoid of the bumblebee, and is found only in bumblebee nests. In some areas in southern England, as much as 80% of nests can be destroyed. Other parasitic larvae are known to prey upon cicadas and leaf hoppers.

In reverse, moths and butterflies may be subject to parasitic wasps and flies, which may lay eggs on the caterpillars, which hatch and feed inside its body, resulting in death. Although, in a form of parasitism called idiobiont, the adult paralyzes the host, so as not to kill it but for it to live as long as possible, for the parasitic larvae to benefit the most. In another form of parasitism, koinobiont, the species live off their hosts while inside (endoparasitic). These parasites live inside the host caterpillar throughout its life cycle, or may affect it later on as an adult. In other orders, koinobionts include flies, a majority of coleopteran, and many hymenopteran parasitoids. Some species may be subject to a variety of parasites, such as the gypsy moth (*Lymantaria dispar*), which is attacked by a series of 13 species, in six different taxa throughout its life cycle.

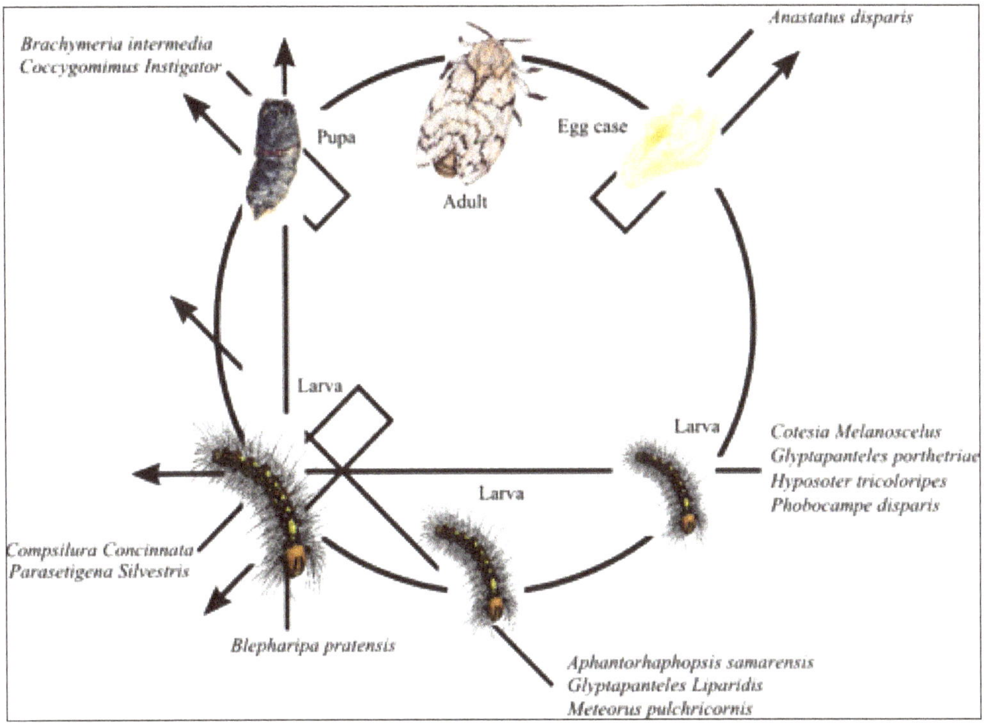

The different parasitoids affecting the gypsy moth (Lymantaria dispar): The stage they affect and eventually kill and its duration are denoted by arrows.

In response to a parsitoid egg or larva in the caterpillar's body, the plasmatocytes, or simply the host's cells can form a multilayered capsule that eventually causes the endoparasite to asphyxiate. The process, called encapsulation, is one of the caterpillar's only means of defense against parasitoids.

Other Biological Interactions

A few species of Lepidoptera are secondary consumers, or predators. These species typically prey upon the eggs of other insects, aphids, scale insects, or ant larvae. Some caterpillars are cannibals, and others prey on caterpillars of other species (e. g. Hawaiian *Eupithecia*). Those of the 15 species in *Eupithecia* that mirror inchworms, are the only known species of butterflies and moths that are ambush predators. Four species are known to eat snails. For example, the Hawaiian caterpillar (*Hyposmocoma molluscivora*) uses silk traps, in a manner similar to that of spiders, to capture certain species of snails (typically Tornatellides).

Larvae of some species of moths in the Tineidae, Gelechioidae, and Noctuidae (family/ superfamily/families, respectively), besides others, feed on detritus, or dead organic material, such as fallen leaves and fruit, fungi, and animal products, and turn it into humus. Well-known species include the cloth moths (*Tineola bisselliella, Tinea pellionella,* and *Trichophaga tapetzella*), which feed on detritus containing keratin, including hair, feathers, cobwebs, bird nests (particularly of domestic pigeons, *Columba livia domestica*) and fruits or vegetables. These species are important to ecosystems as they remove substances that would otherwise take a long time to decompose.

In 2015 it was reported that wasp bracovirus DNA was present in Lepidoptera such as monarch butterflies, silkworms and moths.

BLATTODEA

Blattodea is an order of insects that contains cockroaches and termites. Formerly, the termites were considered a separate order, Isoptera, but genetic and molecular evidence suggests an intimate relationship with the cockroaches, both cockroaches and termites having evolved from a common ancestor. The Blattodea and the mantises (order Mantodea) are now all considered part of the superorder Dictyoptera. Blattodea includes approximately 4,400 species of cockroach in almost 500 genera, and about 3,000 species of termite in around 300 genera.

Termites are pale-coloured, soft-bodied eusocial insects that live in colonies, whereas cockroaches are darker-coloured (often brown), sclerotized, segmented insects. Within the colony, termites have a caste system, with a pair of mature reproductives, the king and the queen, and numerous sterile workers and soldiers. The cockroaches are not colonial but do have a tendency to aggregate and may be considered pre-social, as all adults are capable of breeding. Other similarities between the two groups include various social behaviours, trail following, kin recognition and methods of communication.

Diversity

Over 4000 species of cockroaches are found in every corner of the globe with each continent having its own indigenous species. Most of these are omnivores or detritivores and live in a range of habitats such as among leaf litter, in rotting wood, in thick vegetation, in crevices, in cavities beneath bark, under logs and among debris. Some are arboreal, some live in caves and some are aquatic. A small number of species have taken to living in close proximity to humans in buildings, have been transported around the world by them, and are regarded as pests. Although some species harbour symbionts in their guts which facilitate cellulose digestion, many species also produce enzymes to digest cellulose independent of the symbionts.

Over 3000 species of termite are found in all the continents except Antarctica. The greatest diversity is found in Africa and relatively few species inhabit Europe and North America. They are also detrivores and many species eat wood, having specialised guts with symbiotic protozoa to digest the cellulose. Termites have soft bodies and keep out of sight as far as possible. They can loosely be subdivided into dampwood, drywood and subterranean types. In general, dampwood termites inhabit coniferous forests, drywood termites inhabit hardwood forests and subterranean termites live in a wide variety of habitats.

Characteristics

Termites are eusocial insects that live in colonies. They operate a caste system, there being a king and queen in each colony and many non-reproductive workers. The workers forage for food which they bring back to the colony to feed the reproductives and the developing young. Cockroaches are also social insects but do not live in colonies, and all adults are able to reproduce. Some species form aggregations, others show an inclination to aggregate, and some exhibit parental care of their offspring.

The two groups have striking similarities in behaviour which they likely inherited from their common ancestor. These include an attraction to warm and humid places, thigmotaxis, burrowing, substrate manipulation, hygienic behaviour, food sharing, cannibalism, excretion behaviour, vibrational communication, kin recognition, trail following, allogrooming, care of the brood, cropping of antennae and certain mating behaviours. In some of these behaviours, there are marked similarities between termites and juvenile, but not adult, cockroaches. During the evolution of eusociality, the individuals need to share a desire to group together. Juvenile cockroaches have a tendency to aggregate while adults often compete aggressively with each other for space and resources. Similarly, grooming and being groomed is common in termite colonies but allogrooming is not a behaviour generally engaged in by cockroaches although individuals groom themselves. An exception to this is the cockroach *Cryptocercus*, which seems to be more closely related to the termites than to other cockroaches. Here juveniles groom each other and also groom adults.

Both groups are also affected by their social environments. A single termite, kept alone, has a significantly decreased level of vigour and a shorter lifespan than when two are kept together. An isolated cockroach nymph may grow at less than half the rate of grouped individuals, and has a poorer life expectancy.

Both termites and cockroaches engage in coprophagy, the consumption of fecal pellets. Adult termite workers forage and bring food back to the nest where they pass it to the reproductives and young either by mouth or by anus, providing the whole of their nutritional needs in this manner. Young cockroaches are ineffective foragers, seldom straying from their hiding places, and obtain much of their nourishment from eating the fecal pellets of larger individuals. From these they acquire the microbial flora that helps them to digest their food.

A single cockroach family, the Cryptocercidae, and one primitive species of termite, *Mastotermes darwiniensis*, share such characteristics as the segmental origin of certain female reproductive structures, and the fact that both deposit their eggs in the oothecae that are typical of cockroaches.

Cockroaches

American cockroach.

Arthropods similar to living cockroaches dominated the insect communities of the Carboniferous period. Modern cockroaches radiated from them by the middle of the Mesozoic. This group of insects are nocturnal, only foraging for food and water at night. They are not considered eusocial because their populations are not divided into different caste systems; however, they are still social creatures and can live in groups with over a million individuals. The cockroach is flattened dorsolaterally and is roughly oval with a shield-like plate, the pronotum, covering its thorax and posterior region of the head. The antennae are many-segmented, long and slender, and the mouthparts are adapted for chewing. The forewings are normally leathery and the hind wings membranous. The coxae of the legs are flattened to enable the femurs to fit neatly against them when folded. Cockroaches are hemimetabolous; there is no pupal stage and the nymphs resemble the adults apart from their size and the absence of wings. Female cockroaches produce an egg sac known as an ootheca and can hold anywhere from 12-25 eggs depending upon the species. Some species display parenting behavior, whereas other species have nothing to do with the young. In most species, growth to maturity takes three to four months, but in a few species, the nymph stage can last for several years. The main factors affecting the duration of the nymph stage are seasonal differences, and the amount of nutrients received in the diet.

Chemical Communication

As in most insect species, cockroaches communicate with one another by the release of pheromones. It has also been discovered that cockroaches release hydrocarbons from their body that are transferred through interactions of the antennae. These hydrocarbons can aid in cockroach communication and can even tell whether an individual is a member of its kin or not to prevent inbreeding. Cockroaches that have been isolated in a lab setting have shown extreme behavioral effects and are less stimulated by these hydrocarbons and pheromones, possibly suggesting a group environment is required for development of these communication skills.

Termites

All species of termite are to some degree eusocial, and the members of a colony are differentiated into caste systems. The majority of termite populations consist of the worker caste, which are responsible for foraging, nest building, grooming, and brood care. The soldier caste has one responsibility, which is to protect the nest from predators and other competitors. Soldiers have highly developed mandibles as well as many exocrine glands that can secrete multiple defensive substances harmful to predators.

Termites.

Normally, only the king and queen termite reproduce; the other castes are all sterile. There are two classes of reproductives: Primary reproductives and neotenic reproductives. The primary reproductives class is responsible for colony creation and is characterized by compound eyes, wing marks (spots where wings once were before shedding), and defined sclerotization. Neonetic reproductives can develop from within the colony usually when one of the primary reproductives has died, or can develop in addition to the queen. Neonetic reproductives can experience two different phenotypes, one with wings and one without. If neonetics are winged they will fly away from the parental colony, pair up and form a new colony, but if they are wingless they will remain within the parental colony. The different developmental routes taken by these two morphs are usually dependent upon food availability in the colony, or varying levels of parasitism within the colony. The caste into which any particular nymph will develop begins to become apparent among the late instars; at this time, potential reproductives will begin to show an increase in the size of the gonadal region.

Cathedral termite mounds, Northern Territory.

Termite colonies may be arboreal, mound-like or subterranean, with primitive termites nesting completely inside enclosed structures such as stumps or logs. Nest construction is largely from the termites' own faecal matter, other materials being chewed vegetable fibre, which makes a weak carton-like but waterproof substance, and soil, which makes a strong substance, but which is subject to erosion by water. Aerial nests are connected to the ground by enclosed passageways; the soft-bodied, blind workers of most species live permanently in their protected environments and do not venture into the open air. *Trinervitermes trinervoides* is an exception to this, with workers foraging in small groups on the surface at night, secreting noxious terpenes to deter predators. The nests are complex structures, and tunnels link them to the foraging areas. In Africa, termite mounds can be as large as nine meters tall and thirty meters in diameter, producing an area of increased fertility and creating a small hotspot for biodiversity.

HEMIPTERA

Hemiptera is a major order of insects, collectively known as true bugs, characterized by sucking mouthparts, rather than chewing mouthparts. This is a very successful group of insects, with about 80,000 species, and found in terrestrial and aquatic environments, including the only known marine-pelagic insects. Among groups included in Hemiptera are cicadas, aphids, planthoppers, leafhoppers, shield bugs, assassin bugs, bed bugs, and others. While many feed on sap and other plant material, some feed on animal sources, such as the hemolymph of other arthropods and the blood of mammals and birds.

Hemipterans are vitally important as part of food chains, being consumed by countless vertebrates and invertebrates. And as one of the few and perhaps the most important insect group that has adapted to aquatic environments, they provide an important source of food for fish, and, being voracious predators, help to control mosquitoes during the mosquito's aquatic larval stage. While many hemipterans, such as aphids and scale insects, are significant pests of crops and gardens, some predatory forms serve to control agricultural pests.

The defining feature of hemipterans is their possession of tubular, sucking mouthparts where the mandibles and maxillae are in the form of a proboscis, sheathed within a modified labium to form a "beak" or "rostrum." This structure is capable of piercing tissues (usually plant tissues) and sucking out the liquids, such as sap. Hemipterans do not have chewing mouthparts and any food source that is going to be ingested has to be in liquid form.

Essentially, all hemipterans have a piercing rostrum or labium that is in distinct segments. The sucking mouthparts are formed from tubular structures called stylets that are modified from their use in other insects groups, where they surround the jaws. In hemipterans, the stylets of the mandible, or lower jaw, are serrated and surround the stylets of the maxilla, or upper jaw, which are smooth and are longer. The serrated mandibular stylets help to cut into the feeding substrate, while the smooth maxillary stylets are then pushed further into the substrate to such the liquid into the insect. In addition, these modified structures include a pump for sucking, a precibarium for tasting, and muscles to operate the pump. And the smooth maxillary stylets release saliva with enzymes that help to partially digest the food source so that it can be in the form of a "soup" for sucking into the insect.

The antennae in Hemiptera are typically five-segmented, although they can still be quite long, and the tarsi of the legs are three-segmented or shorter.

Although hemipterans vary widely in their overall form, their mouthparts, formed into a rostrum, are quite distinctive; the only orders with mouthparts modified in a similar manner are the Thysanoptera and some Phthiraptera, and these are generally easy to recognize as non-hemipteran for other reasons. Aside from the mouthparts, various insects can be confused with hemipterans, including cockroachesand psocids, both of which have longer many-segmented antennae, and some beetles, but these have fully-hardened forewings, which do not overlap.

Hemipterans range in size from 1 millimeter to around 15 centimeters (CSIRO).

Life Cycle and Ecology

Hemipterans are hemimetabolous insects, meaning that they do not undergo complete metamorphosis or "holometabolism," in which the larvae differ markedly from the adults, but rather the larvae resemble the adults somewhat. That is, the young have compound eyes, developed legs, and wing stubs visible on the outside, but the juvenile forms are smaller and, if the adult has wings, lack wings. In this mode of development, there are three distinct stages: the egg, nymph, and the adult stage, or imago. These groups go through gradual changes; there is no pupal stage. The final transformation involves little more than the development of functional wings (if they are present at all) and functioning sexual organs. Hemiptera is the largest insect order that is hemimetabolous; the orders with more species all have a pupal stage (Lepidoptera, Coleoptera, Diptera, and Hymenoptera).

Many aphids are parthenogenetic during part of the life cycle, such that females can produce unfertilized eggs, which are clones of themselves.

Most hemipterans are phytophagous, feeding on plant sap. Such plant feeders include aphids, scale insects, cicadas, and others. Most of the remainder are predatory, feeding on other insects, or even small vertebrates. A few, however, are parasites, feeding on the blood of larger animals. These include bedbugs and the kissing bugs of the family Reduviidae, which can transmit potentially deadly Trypanosoma infections.

Pondskaters Gerris najasmating.

Several families of Hemiptera are water bugs, adapted to an aquatic lifestyle, such as the water boatmen and water scorpions. They are mostly predatory, and have legs adapted as paddles to help the animal move through the water. The "pondskaters" or "water striders" of the family Gerridae are also associated with water, but use the surface tension of standing water to keep them above the surface; they include the genus Halobates which is the only group of insects to be truly marine-pelagic, found on patches of flotsam in the open ocean.

Classification

The present members of the order Hemiptera were historically placed into two orders, Homoptera and Heteroptera/Hemiptera, based on the differences in wing structure and the position of the rostrum. These two orders were then combined into the single order Hemiptera by many authorities, with Homoptera and Heteroptera classified as suborders. The Hemiptera order is presently more typically divided into four or more suborders, after it was established that the families grouped together as "Homoptera" are not as closely related as had previously been thought.

Four orders are commonly recognized: Auchenorrhyncha, Coleorrhyncha, Heteroptera, and Sternorrhyncha. Auchenorrhyncha contains the cicadas, leafhoppers, treehoppers, planthoppers, and froghoppers. The 12,500 species in the suborder Sternorrhyncha are the aphids, whiteflies, and scale insects. The suborder Coleorrhyncha (comprising the single family Peloridiidae), contains fewer than 30 species of Gondwana-distributed bugs, and is sometimes grouped with the Heteroptera (to form the suborder Prosorrhyncha). Heteroptera itself is a group of 25,000 species of relatively large bugs, including the shield bugs, seed bugs, assassin bugs, flower bugs, and the water bugs.

The closest relatives of hemipterans are the thrips and lice, which collectively form the "Hemipteroid Assemblage" within the Exopterygota subclass of the Class Insecta.

Economic Significance

Many species of Hemiptera are significant pests of crops and gardens, including many species of aphid (such as whitefly, greenfly, and blackfly) and various scale insects. Thc later includes the cottony cushion scale, a pest whose infestation of American citrus crops sparked one of the earliest biological pest control programs, when the Australian beetle Rodolia cardinalis was introduced as a natural enemy of the scale insect.

Conversely, some predatory hemipterans are themselves biological pest control agents, such as various nabids (Mahr) and even some members of families that are primarily phytophagous, such as the genus Geocoris in the family Lygaeidae (Hagler). Other hemipterans have positive uses, such as in the production of the dyestuffs cochineal and crimson, or shellac.

References

- Class-insecta: iaszoology.com, Retrieved 23 July, 2020
- Bouchard, P.; Bousquet, Y.; Davies, A.; Alonso-Zarazaga, M.; Lawrence, J.; Lyal, C.; Newton, A.; Reid, C.; Schmitt, M.; Ślipiński, A.; Smith, A. (2011). "Family-group names in Coleoptera (Insecta)". ZooKeys (88): 1–972. doi:10.3897/zookeys.88.807. PMC 3088472. PMID 21594053
- Resh, Vincent H.; Cardé, Ring T. (2009). Encyclopedia of Insects. Academic Press. pp. 284–297. ISBN 978-0-08-092090-0
- Hemiptera: newworldencyclopedia.org, Retrieved 20 August, 2020
- Janke, Axel; Klopfstein, Seraina; Vilhelmsen, Lars; Heraty, John M.; Sharkey, Michael; Ronquist, Fredrik (2013). "The Hymenopteran Tree of Life: Evidence from Protein-Coding Genes and Objectively Aligned Ribosomal Data". PLoS ONE. 8 (8): e69344. Bibcode:2013PLoSO...869344K. doi:10.1371/journal.pone.0069344. PMC 3732274. PMID 23936325
- Beccaloni, G. W.; Eggleton, P. (2011). "Order Blattodea Brunner von Wattenwyl, 1882" (PDF). In Zhang, Z.-Q. (ed.). Animal biodiversity: An outline of higher-level classification and survey of taxonomic richness. Zootaxa. 3148. Magnolia Press. pp. 199–200. doi:10.11646/zootaxa.3148.1.37. ISBN 978-1-86977-850-7. ISSN 1175-5334

External Anatomy

CHAPTER 3

External anatomy of insects comprises of three major body regions – head, thorax and abdomen. Head has five to seven fused segments with eyes, antennae and mouthparts. Thorax can be divided into three segments and abdomen comprises eleven or fewer segments. This chapter discusses the external anatomy of insects in detail.

Insects are segmented invertebrates that possess the articulated external skeleton (exoskeleton) characteristic of all arthropods. Groups are differentiated by various modifications of the exoskeleton and the appendages — for example, the Hexapoda to which the Insecta belong is characterized by having six-legged adults. Many anatomical features of the appendages, especially of the mouthparts, legs, wings, and abdominal apex, are important in recognizing the higher groups within the hexapods, including insect orders, families, and genera. Differences between species frequently are indicated by less obvious anatomical differences. Furthermore, the biomechanical analysis of morphology (e.g. studying how insects fly or feed) depends on a thorough knowledge of structural features. Clearly, an understanding of external anatomy is necessary to interpret and appreciate the functions of the various insect designs and to allow identification of insects and their hexapod relatives.

First some basic classification and terminology needs to be explained. Adult insects normally have wings (most of the Pterygota), the structure of which may diagnose orders, but there is a group of primitively wingless insects (the "apterygotes"). Within the Insecta, three major patterns of development can be recognized. Apterygotes (and non-insect hexapods) develop to adulthood with little change in body form (ametaboly), except for sexual maturation through development of gonads and genitalia. All other insects either have a gradual change in body form (hemimetaboly) with external wing buds getting larger at each molt, or an abrupt change from a wingless immature insect to winged adult stage via a pupal stage (holometaboly). Immature stages of hemimetabolous insects are generally called nymphs, whereas those of holometabolous insects are referred to as larvae.

Anatomical structures of different taxa are homologous if they share an evolutionary origin, i.e. if the genetic basis is inherited from an ancestor common to them both. For instance, the wings of all insects are believed to be homologous; this means that wings originated once. Homology of structures generally is inferred by comparison of similarity in ontogeny (development from egg to adult), composition (size and detailed appearance), and position (on the same segment and same relative location on that segment). The homology of insect wings is demonstrated by similarities in venation and articulation – the wings of all insects can be derived from the same basic pattern or groundplan. Sometimes association with other structures of known homologies is helpful in establishing the homology of a structure of uncertain origin. Another sort of homology, called serial homology, refers to corresponding structures on different segments of an individual insect. Thus, the appendages of each body segment are serially homologous, although in living insects those on the head (antennae and mouthparts) are very different in appearance from those on the thorax (walking legs) and abdomen (genitalia and cerci).

CUTICLE

The cuticle forms the major part of the integument of the Arthropoda. It includes most of the material of the exoskeleton of the insects, Crustacea, Arachnida, and Myriapoda.

Cuticles of some armoured beetles are particularly rigidly sclerotised.

Honeybee larvae have flexible but delicate unsclerotised cuticles.

Morphology

In arthropods, the integument, the external "skin", or "shell", is the product of a single layer of ectodermal epithelium. That layer is attached to the external or distal surface of the deepest layer, the non-cellular internal membrane of the integument. That non-cellular membrane is called the basement membrane. The layer of epithelium on the basement membrane produces the cuticle, which begins as a tough, flexible layer of chitin. Such thin, flexible chitin is the major structural part of the integument where flexibility is necessary, such as in bodily parts that must stretch to contain accumulated liquids, or that form joints between rigid parts of the exoskeleton. In other parts of the cuticle the function of the integument demands more rigid materials, such as armoured regions or the biting parts of the jaws, or where the exoskeleton forms the tubular limbs of most Arthropoda. To achieve such rigidity the outer chitin layer of the cuticle is impregnated, thickened, and reinforced with harder, more brittle materials such as sclerotinised proteins or calcite. This main chitinous layer of the cuticle is called the *procuticle*.

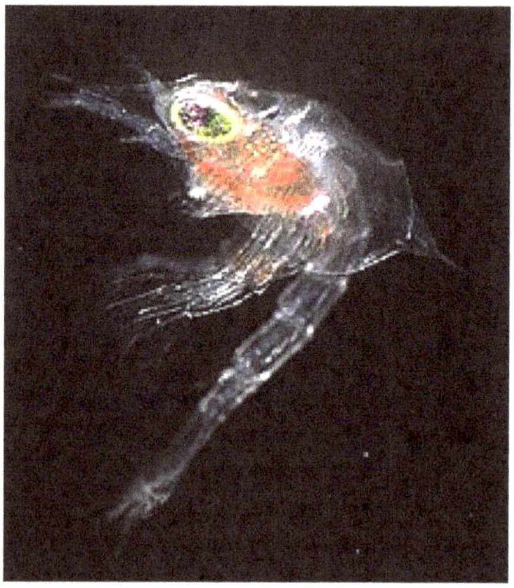

This Zoea-stage larva is hardly recognisable as a crab, but each time it sheds its cuticle it remodels itself, eventually taking on its final crab form.

The procuticle in most land Arthropoda, particularly insects, is covered externally with a thin, waxy, water-resistant outer layer containing no chitin. That outer layer is the *epicuticle*, and it is much thinner than the procuticle. The chitinous procuticle is formed of an outer *exocuticle* and the inner *endocuticle*, and between the *exocuticle* and *endocuticle* there may be another layer called *mesocuticle* which has distinctive staining properties. The tough and flexible endocuticle is a laminated structure of layers of interwoven fibrous chitin and protein molecules, while the exocuticle is the layer in which any major thickening, armouring and biomineralization occurs. Biomineralization with calcite is particularly common in Crustacea, whereas sclerotization particularly occurs in insects. The exocuticle is greatly reduced in many soft-bodied insects, especially in the larval stages such as caterpillars and the larvae of parasitoidal Hymenoptera.

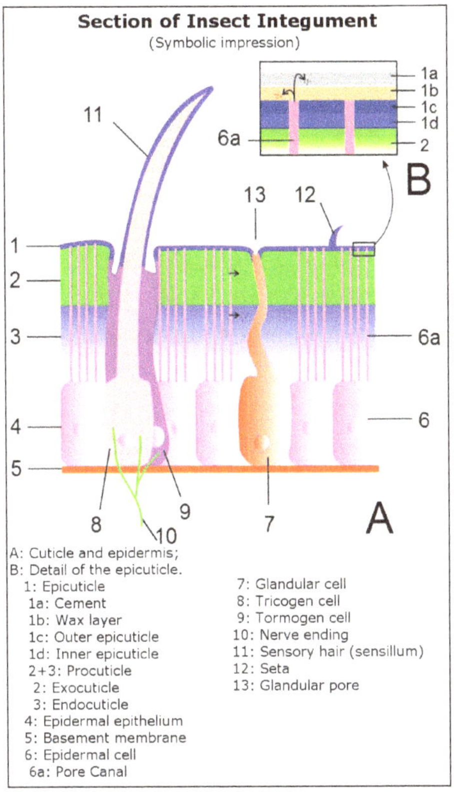

Hardened plates in the exoskeleton are called sclerites. Sclerites may be simple protective armour, but also may form mechanical components of the exoskeleton, such as in the legs, joints, fins or wings. In the typical body segment of an insect or many other Arthropoda, there are four principal regions. The dorsal region is the tergum; if the tergum bears any sclerites, those are called *tergites*. The ventral region is called the sternum, which commonly bears sternites. The two lateral regions are called the pleura (singular pleurum) and any sclerites they bear are called pleurites.

This fully-grown robber crab has tough fabric forming its joints, delicate biomineralized cuticle over its sensory antennae, optic-quality over its eyes, and strong, calcite-reinforced chitin armouring its body and legs; its pincers can break into coconuts.

Within entomology, the term *glabrous* is used to refer to those parts of an insect's body lacking in setae (bristles) or scales.

Chemical Composition

Chemically, chitin is a long-chain polymer of a N-acetylglucosamine, which is a derivative of glucose. The polymer bonds between the glucose units are $\beta(1\rightarrow4)$ links, the same as in cellulose.

In its unmodified form, chitin is translucent, pliable, resilient and tough. In arthropods and other organisms however, it generally is a component of a complex matrix of materials. It practically always is associated with protein molecules that often are in a more or less sclerotised state, stiffened or hardened by cross-linking and by linkage to other molecules in the matrix. In some groups of animals, most conspicuously the Crustacea, the matrix is greatly enriched with, or even dominated by, hard minerals, usually calcite or similar carbonates that form much of the exoskeleton. In some organisms the mineral content may exceed 95%. The role of the chitin and proteins in such structures is more than just holding the crystals together; the crystal structure itself is so affected as to prevent the propagation of cracks under stress, leading to remarkable strength. The process of formation of such mineral-rich matrices is called biomineralization.

The difference between the unmodified and modified forms of chitinous arthropodan exoskeletons can be seen by comparing the body wall of say a bee larva, in which modification is minimal, to any armoured species of beetle, or the fangs of a spider. In both those examples there is heavy modification by sclerotisation. Again, contrasting strongly with both unmodified organic material such as largely pure chitin, and with sclerotised chitin and proteins, consider the integument of a heavily armoured crab, in which there is a very high degree of modification by biomineralization.

Moulting

The integument, and in particular the cuticle, is the product of the single layer of columnar or cuboidal epithelial cells attached to the basement membrane. The cuticle provides muscular support and acts as a protective shield as the insect develops, but it is not in itself cellular, so once established it cannot grow and offers little scope for maintenance, renewal or increase in size as the animal grows. Therefore, the animal periodically sheds the external part of the cuticle in a process called moulting or ecdysis. As the time for moulting approaches, as much as possible of the exocuticle material is internally digested by enzymatic action and reabsorbed through the epithelial layer. As its base gets dissolved away, the old cuticle begins to separate from the epidermis in a process called apolysis.

Early in the process of apolysis the epithelial cells release enzymatic moulting fluid between the old cuticle and the epidermis. The enzymes partly digest the endocuticle and the epidermis absorbs the digested material for the animal to assimilate. Much of that digested material is re-used to build the new cuticle. Once the new cuticle has formed sufficiently, the animal splits the remaining parts of the old integument along built-in lines of weakness and sheds them in the visible process of ecdysis, generally shedding and discarding the epicuticle and the reduced exocuticle, though some species carry them along for camouflage or protection. The shed portions are called the exuviae. The animal then expands its body by swallowing liquid or gas and in the process it stretches the new integument to its proper size and shape. The new integument still is soft and usually is pale, and it is said to be teneral or callow. It then undergoes a hardening and pigmentation process that might take anything from several minutes to several days, depending on the nature of the animal and the circumstances.

Cuticle Design

Basement Membrane

The basement membrane is connected to the layer of epidermal cells. The membrane consists of collagen fibres, as has been shown for Locusta migratoria, and for the beetle, Hydrocyrius columbiae. It has been hypothesised that the membrane derives from plasmatocytes in the haemolymph.

Epicuticle

The epicuticle is the very thin, outermost layer of the cuticle. However, it is complex and consists of several sublayers, namely 1) cement sublayer, 2) wax sublayer, 3) outer epicuticle, and 4) inner epicuticle. Each sublayer is of unique composition and properties. This composite structure does not contain chitin, and overlies the chitin-bearing procuticle. The epicuticle covers the surface of an insect. However, some exceptions are known: it is missing, for example, in chemoreceptors and midgut.

Functions of the epicuticle are water homeostasis, setting limits on the dimensions of an exoskeleton, and transport of moulting fluids and epidermal secretions. The epicuticle may also serve as a chemical reservoir for the storage of metabolic waste products, or for the brief retention of juvenile hormones or defensive secretions. The high extension ability of the recently ecdysed cuticle, and specialised extending cuticles, such as, in fluid-sucking insects, is a result of the mechanical properties of the epicuticle.

The outermost layer is a cement sublayer that is subtended by a wax layer. The cement layer is the combined secretory product of dermal glands. It consists of proteins and lipids stabilised by various polyphenolic substances. The distribution and thickness of this layer varies even in the same insect species. Some insects, such as the honey-bee, completely lack the cement sublayer. The primary function of the sublayer is to protect the lipid monolayer and outer epicuticle from possible abrasion, or impact damage.

HEAD

The head in most insects is enclosed in a hard, heavily sclerotized, exoskeletal head capsule'. The main exception is in those species whose larvae are not fully sclerotised, mainly some holometabola; but even most unsclerotised or weakly sclerotised larvae tend to have well sclerotised head capsules, for example the larvae of Coleoptera and Hymenoptera. The larvae of Cyclorrhapha however, tend to have hardly any head capsule at all.

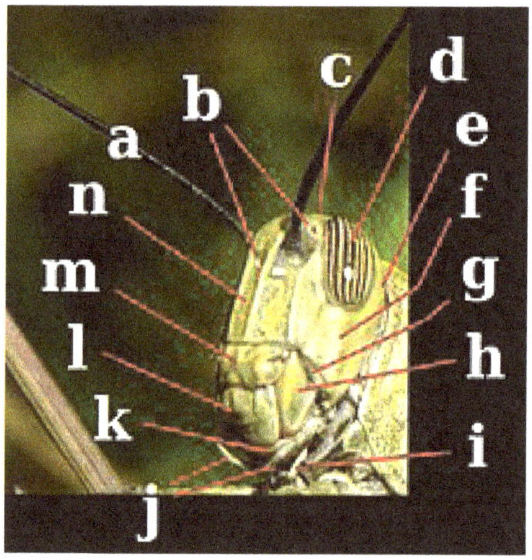

Head of Orthoptera, Acrididae. a:Antenna, b:Ocelli, c:Vertex, d:Compound eye, e:Occiput, f:Gena, g:Pleurostoma, h:Mandible, i:Labial palp, j:Maxillary palps, k:Maxilla, l:Labrum, m:Clypeus, n:Frons.

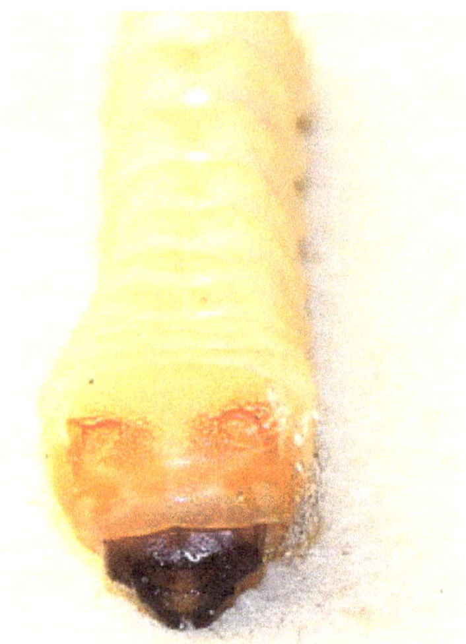

Larva of beetle, family Cerambycidae, showing sclerotised epicranium; rest of body hardly sclerotized.

The head capsule bears most of the main sensory organs, including the antennae, ocelli, and the compound eyes. It also bears the mouthparts. In the adult insect the head capsule is apparently unsegmented, though embryological studies show it to consist of six segments that bear the paired head appendages, including the mouthparts, each pair on a specific segment. Each such pair occupies one segment, though not all segments in modern insects bear any visible appendages.

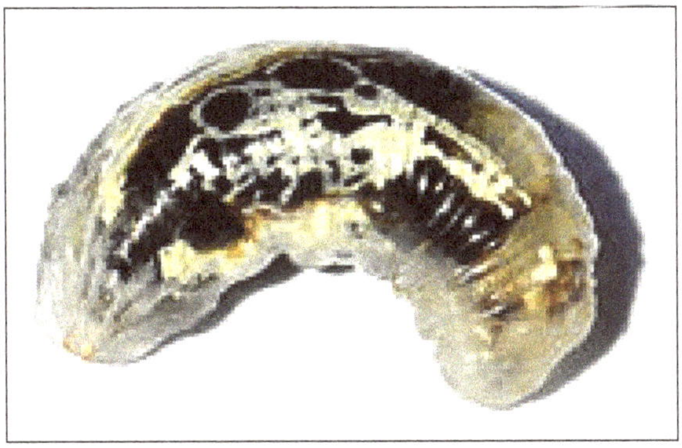

Larva of Syrphid fly, member of Cyclorrhapha, without epicranium, almost without sclerotisation apart from its jaws.

Of all the insect orders, Orthoptera most conveniently display the greatest variety of features found in the heads of insects, including the sutures and sclerites. Here, the vertex, or the apex (dorsal region), is situated between the compound eyes for insects with hypognathous and opisthognathous heads. In prognathous insects, the vertex is not found between the compound eyes, but rather, where the ocelli are normally found. This is because the primary axis of the head is rotated 90° to become parallel to the primary axis of the body. In some species, this region is modified and assumes a different name.

The ecdysial suture is made of the coronal, frontal, and epicranial sutures plus the ecdysial and cleavage lines, which vary among different species of insects. The ecdysial suture is longitudinally placed on the vertex and separates the epicranial halves of the head to the left and right sides. Depending on the insect, the suture may come in different shapes: like either a Y, U, or V. Those diverging lines that make up the ecdysial suture are called the frontal or frontogenal sutures. Not all species of insects have frontal sutures, but in those that do, the sutures split open during ecdysis, which helps provide an opening for the new instar to emerge from the integument.

The frons is that part of the head capsule that lies ventrad or anteriad of the vertex. The frons varies in size relative to the insect, and in many species the definition of its borders is arbitrary, even in some insect taxa that have well-defined head capsules. In most species, though, the frons is bordered at its anterior by the frontoclypeal or epistomal sulcus above the clypeus. Laterally it is limited by the fronto-genal sulcus, if present, and the boundary with the vertex, by the ecdysial cleavage line, if it is visible. If there is a median ocellus, it generally is on the frons, though in some insects such as many Hymenoptera, all three ocelli appear on the vertex. A more formal definition is that it is the sclerite from which the pharyngeal dilator muscles arise, but in many contexts that too, is not helpful. In the anatomy of some taxa, such as many Cicadomorpha, the front of the head is fairly clearly distinguished and tends to be broad and sub-vertical; that median area commonly is taken to be the frons.

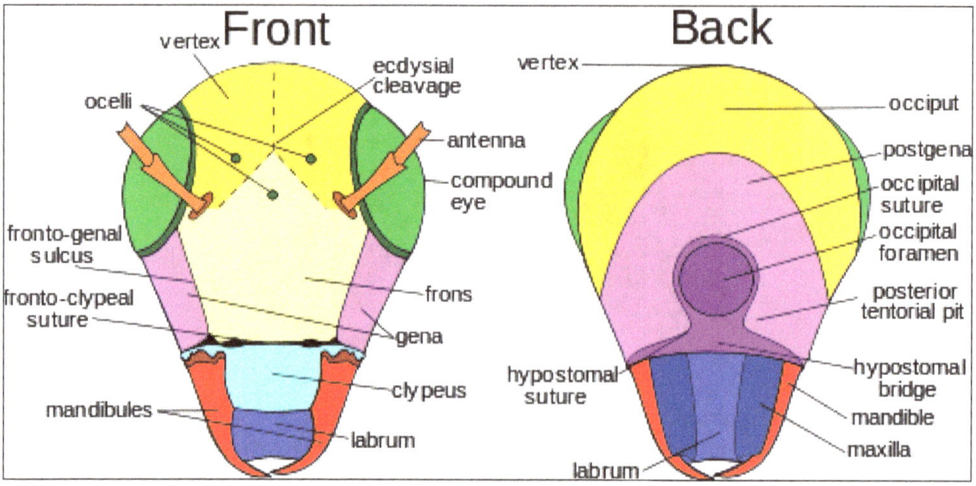

The clypeus is a sclerite between the face and labrum, which is dorsally separated from the frons by the frontoclypeal suture in primitive insects. The clypeogenal suture laterally demarcates the clypeus, with the clypeus ventrally separated from the labrum by the clypeolabral suture. The clypeus differs in shape and size, such as species of Lepidoptera with a large clypeus with elongated mouthparts. The cheek or gena forms the sclerotized area on each side of the head below the compound eyes extending to the gular suture. Like many of the other parts making up the insect's head, the gena varies among species, with its boundaries difficult to establish. For example, in dragonflies and damselflies, it is between the compound eyes, clypeus, and mouthparts. The postgena is the area immediately posteriad, or posterior or lower on the gena of pterygote insects, and forms the lateral and ventral parts of the occipital arch. The occipital arch is a narrow band forming the posterior edge of the head capsule arching dorsally over the foramen. The subgenal area is usually narrow, located above the mouthparts; this area also includes the hypostoma and pleurostoma. The vertex extends anteriorly above the bases of the antennae as a prominent, pointed, concave rostrum. The posterior wall of the head capsule is penetrated by a large aperture, the foramen. Through it pass the organ systems, such as nerve cord, esophagus, salivary ducts, and musculature, connecting the head with the thorax.

On the posterior aspect of the head are the occiput, postgena, occipital foramen, posterior tentorial pit, gula, postgenal bridge, hypostomal suture and bridge, and the mandibles, labium, and maxilla. The occipital suture is well founded in species of Orthoptera, but not so much in other orders. Where found, the occipital suture is the arched, horseshoe-shaped groove on the back of the head, ending at the posterior of each mandible. The postoccipital suture is a landmark on the posterior surface of the head, and is typically near the occipital foremen. In pterygotes, the postocciput forms the extreme posterior, often U-shaped, which forms the rim of the head extending to the postoccipital suture. In pterygotes, such as those of Orthoptera, the occipital foramen and the mouth are not separated. The three types of occipital closures, or points under the occipital foramen that separate the two lower halves of the postgena, are: the hypostomal bridge, the postgenal bridge, and the gula. The hypostomal bridge is usually found in insects with hypognathous orientation. The postgenal bridge is found in the adults of species of higher Diptera and aculeate Hymenoptera, while the gula is found on some Coleoptera, Neuroptera, and Isoptera, which typically display prognathous-oriented mouthpartso.

Compound Eyes and Ocelli

Most insects have one pair of large, prominent compound eyes composed of units called ommatidia (ommatidium, singular), possibly up to 30,000 in a single compound eye of, for example, large dragonflies. This type of eye gives less resolution than eyes found in vertebrates, but it gives acute perception of movement and usually possesses UV- and green sensitiity and may have additional sensitivity peaks in other regions of the visual spectrum. Often an ability to detect the E-vector of polarized light exists polarization of light. There can also be an additional two or three ocelli, which help detect low light or small changes in light intensity. The image perceived is a combination of inputs from the numerous ommatidia, located on a convex surface, thus pointing in slightly different directions. Compared with simple eyes, compound eyes possess very large view angles and better acuity than the insect's dorsal ocelli, but some stemmatal (= larval eyes), for example those of sawfly larvae (Tenthredinidae) with an acuity 4 degrees and very high polarization sensitivity, match the performance of compound eyes.

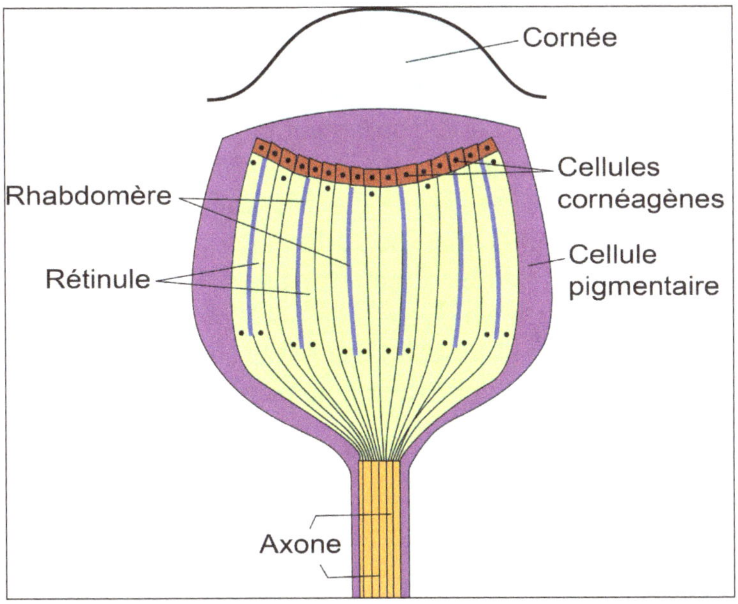

Ocellus cross-section.

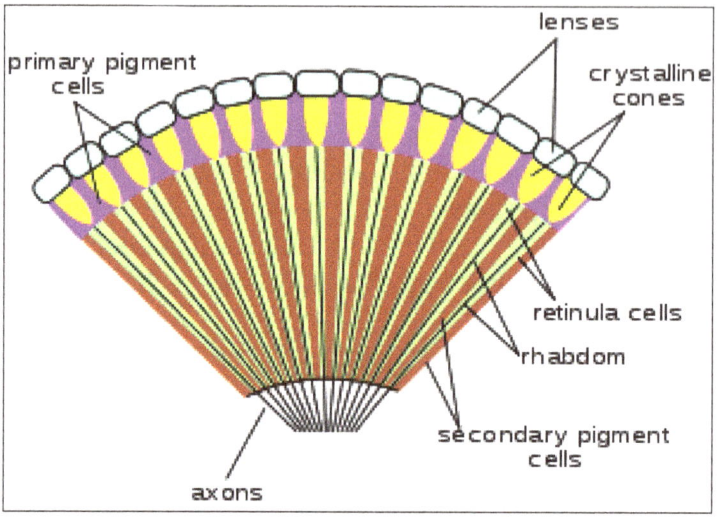

Compound eye cross-section.

Because the individual lenses are so small, the effects of diffraction impose a limit on the possible resolution that can be obtained (assuming they do not function as phased arrays). This can only be countered by increasing lens size and number. To see with a resolution comparable to our simple eyes, humans would require compound eyes that would each reach the size of their heads. Compound eyes fall into two groups: Apposition eyes, which form multiple inverted images, and superposition eyes, which form a single erect image. Compound eyes grow at their margins by the addition of new ommatidia.

Piercing-sucking

Mouthparts can have multiple functions. Some insects combine piercing parts along with sponging ones which are then used to pierce through tissues of plants and animals. Female mosquitoes feed on blood (hemophagous) making them disease vectors. The mosquito mouthparts consist of the proboscis, paired mandibles and maxillae. The maxillae form needle-like structures, called stylets, which are enclosed by the labium. When mosquito bites, maxillae penetrate the skin and anchor the mouthparts, thus allowing other parts to be inserted. The sheath-like labium slides back, and the remaining mouthparts pass through its tip and into the tissue. Then, through the hypopharynx, the mosquito injects saliva, which contains anticoagulants to stop the blood from clotting. And finally, the labrum (upper lip) is used to suck up the blood. Species of the genus *Anopheles* are characterized by their long palpi (two parts with widening end), almost reaching the end of labrum.

Examples of piercing mouthparts:

- Aedes aegypti.
- Mosquito.

- Cimex lectularius.
- Flea.
- Horsefly (female).
- Tick (*Ixodes ricinus*), which is not an insect, but an arachnid, given for comparison.

Siphoning

The proboscis is formed from maxillary galeae and is adaption found in some insects for sucking. The muscles of the cibarium or pharynx are strongly developed and form the pump. In Hemiptera and many Diptera, which feed on fluids within plants or animals, some components of the mouthparts are modified for piercing, and the elongated structures are called stylets. The combined tubular structures are referred to as the proboscis, although specialized terminology is used in some groups.

In species of Lepidoptera, it consists of two tubes held together by hooks and separable for cleaning. Each tube is inwardly concave, thus forming a central tube through which moisture is sucked. Suction is effected through the contraction and expansion of a sac in the head. The proboscis is coiled under the head when the insect is at rest, and is extended only when feeding. The maxillary palpi are reduced or even vestigial. They are conspicuous and five-segmented in some of the more basal families, and are often folded. The shape and dimensions of the proboscis have evolved to give different species wider and therefore more advantageous diets. There is an allometric scaling relationship between body mass of Lepidoptera and length of proboscis from which an interesting adaptive departure is the unusually long-tongued hawk moth *Xanthopan morganii praedicta*. Charles Darwin predicted the existence and proboscis length of this moth before its discovery based on his knowledge of the long-spurred Madagascan star orchid *Angraecum sesquipedale*.

Sponging

The mouthparts of insects that feed on fluids are modified in various ways to form a tube through which liquid can be drawn into the mouth and usually another through which saliva passes. The muscles of the cibarium or pharynx are strongly developed to form a pump. In nonbiting flies, the mandibles are absent and other structures are reduced; the labial palps have become modified to form the labellum, and the maxillary palps are present, although sometimes short. In Brachycera, the labellum is especially prominent and used for sponging liquid or semiliquid food. The labella are a complex structure consisting of many grooves, called pseudotracheae, which sop up liquids. Salivary secretions from the labella assist in dissolving and collecting food particles so they can be more easily taken up by the pseudotracheae; this is thought to occur by capillary action. The liquid food is then drawn up from the pseudotracheae through the food channel into the esophagus.

The mouthparts of bees are of a chewing and lapping-sucking type. Lapping is a mode of feeding in which liquid or semiliquid food adhering to a protrusible organ, or "tongue", is transferred from substrate to mouth. In the honey bee (Hymenoptera: Apidae: *Apis mellifera*), the elongated and fused labial glossae form a hairy tongue, which is surrounded by the maxillary galeae and the labial palps to form a tubular proboscis containing a food canal. In feeding, the tongue is dipped into the nectar or honey, which adheres to the hairs, and then is retracted so the adhering liquid is carried into the space between the galeae and labial palps. This back-and-forth glossal movement occurs repeatedly. Movement of liquid to the mouth apparently results from the action of the cibarial pump, facilitated by each retraction of the tongue pushing liquid up the food canal.

Antenna

Large antennae on a longhorn beetle.

Antennae (singular: antenna), sometimes referred to as "feelers", are paired appendages used for sensing in arthropods.

Antennae are connected to the first one or two segments of the arthropod head. They vary widely in form but are always made of one or more jointed segments. While they are typically sensory organs, the exact nature of what they sense and how they sense it is not the same in all groups. Functions may variously include sensing touch, air motion, heat, vibration (sound), and especially smell or taste. Antennae are sometimes modified for other purposes, such as mating, brooding, swimming, and even anchoring the arthropod to a substrate. Larval arthropods have antennae that differ from those of the adult. Many crustaceans, for example, have free-swimming larvae that use their antennae for swimming. Antennae can also locate other group members if the insect lives in a group, like the ant. The common ancestor of all arthropods likely had one pair of uniramous (unbranched) antenna-like structures, followed by one or more pairs of biramous (having two major branches) leg-like structures, as seen in some modern crustaceans and fossil trilobites. Except for the chelicerates and proturans, which have none, all non-crustacean arthropods have a single pair of antennae.

Crustaceans

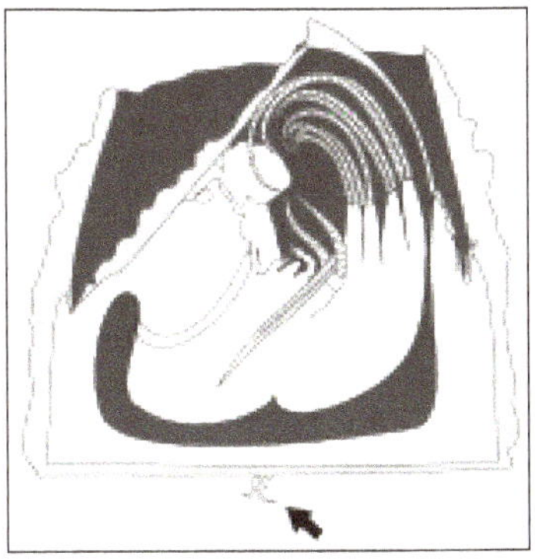

Cutaway diagram of a barnacle, with antennae highlighted by arrow.

Crustaceans bear two pairs of antennae. The pair attached to the first segment of the head are called primary antennae or antennules. This pair is generally uniramous, but is biramous in crabs and lobsters and remipedes. The pair attached to the second segment are called secondary antennae or simply antennae. The second antennae are plesiomorphically biramous, but many species later evolved uniramous pairs. The second antennae may be significantly reduced (e.g. remipedes), apparently absent (e.g. barnacles), or modified to such an extent that they no longer resemble antennae (e.g. spiny lobsters and slipper lobsters).

The subdivisions of crustacean antennae have many names, including flagellomeres (a shared term with insects), annuli, articles, and segments. The terminal ends of crustacean antennae have two major categorizations: segmented and flagellate. An antenna is considered segmented if each of the annuli is separate from those around it and has individual muscle attachments. Flagellate antennae, on the other hand, have muscle attachments only around the base, acting as a hinge for the flagellum—a flexible string of annuli with no muscle attachment.

There are several notable non-sensory uses of antennae in crustaceans. Many crustaceans have a mobile larval stage called a nauplius, which is characterized by its use of antennae for swimming. Barnacles, a highly modified crustacean, use their antennae to attach to rocks and other surfaces.

The large flattened plates in front of the eyes of a slipper lobster are the modified second antennae.

The crab *Cancer pagurus*, showing its reduced antennae.

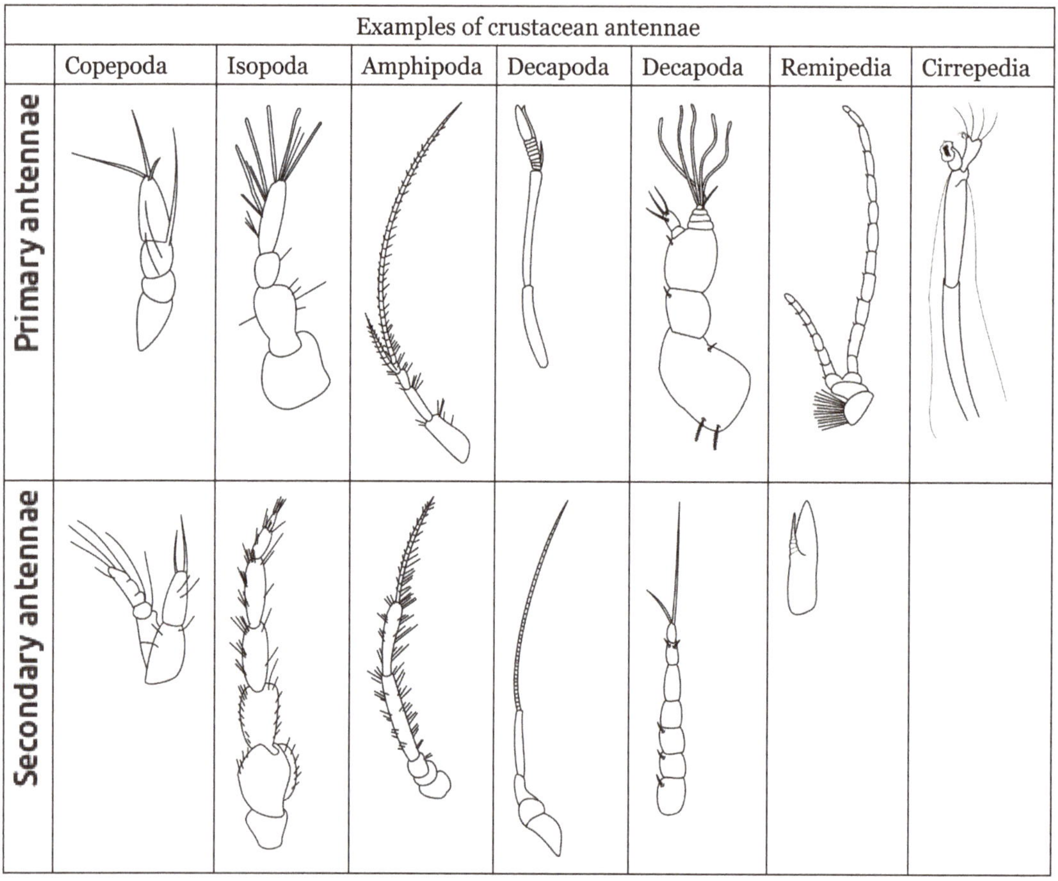

Insects

Insects evolved from prehistoric crustaceans, and they have secondary antennae like crustaceans, but not primary antennae. Antennae are the primary olfactory sensors of insects and are accordingly well-equipped with a wide variety of sensilla (singular: *sensillum*). Paired, mobile, and segmented, they are located between the eyes on the forehead. Embryologically, they represent the appendages of the second head segment.

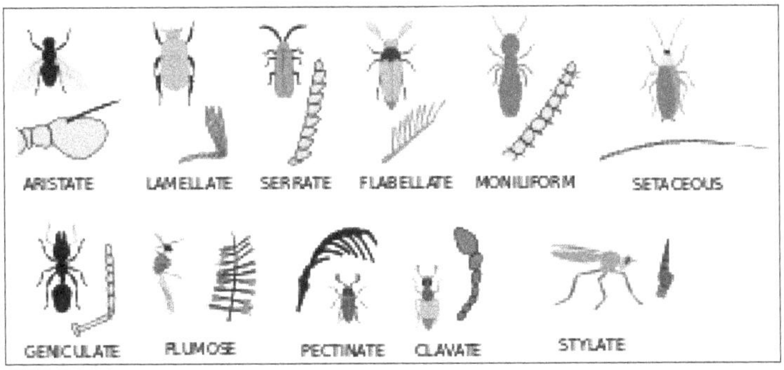

Terms used to describe shapes of insect antennae.

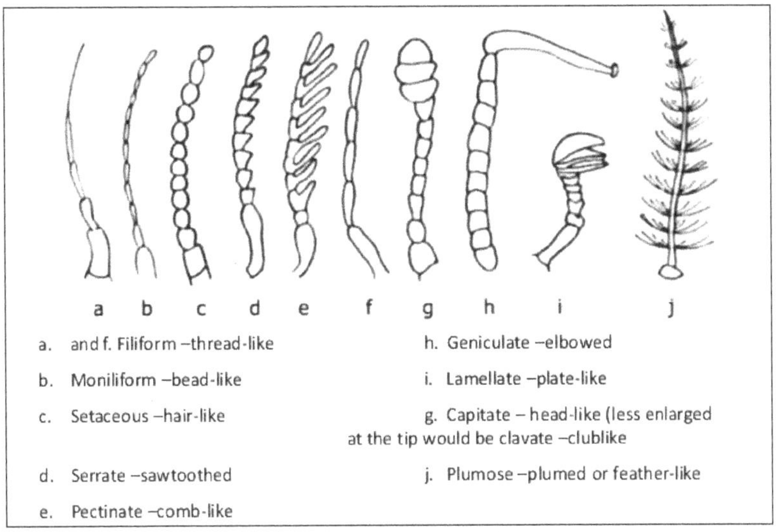

Antennal shape in the Lepidoptera.

All insects have antennae, however they may be greatly reduced in the larval forms. Amongst the non-insect classes of the Hexapoda, both Collembola and Diplura have antenna, but Protura do not.

Antennal fibrillae play an important role in *Culex pipiens* mating practices. The erection of these fibrillae is considered to be the first stage in reproduction. These fibrillae serve different functions across the sexes. As antennal fibrillae are used by female *C. pipiens* to locate hosts to feed on, male *C. pipiens* utilize them to locate female mates.

Structure

The three basic segments of the typical insect antenna are the scape or *scapus* (base), the pedicel or *pedicellus* (stem), and finally the flagellum, which often comprises many units known as flagellomeres. The pedicel (the second segment) contains the Johnston's organ which is a collection of sensory cells.

The scape is mounted in a socket in a more or less ring-shaped sclerotised region called the torulus, often a raised portion of the insect's head capsule. The socket is closed off by the membrane into which the base of the scape is set. However, the antenna does not hang free on the membrane, but pivots on a rigidly sprung projection from the rim of the torulus. That projection on which the antenna pivots is called the antennifer. The whole structure enables the insect to move the antenna as a whole by applying internal muscles connected to the scape. The pedicel is flexibly connected to the distal end of the scape and its movements in turn can be controlled by muscular connections between the scape and pedicel. The number of flagellomeres can vary greatly between insect species, and often is of diagnostic importance.

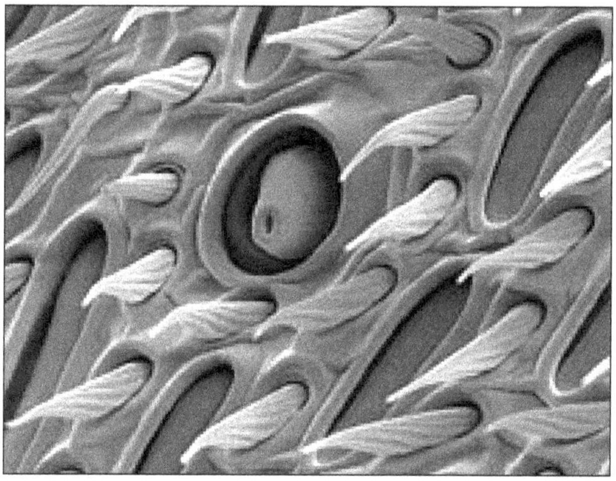

Electron micrograph of antenna surface detail of a wasp *(Vespula vulgaris)*.

True flagellomeres are connected by membranous linkage that permits movement, though the flagellum of "true" insects does not have any intrinsic muscles. Some other Arthropoda do however have intrinsic muscles throughout the flagellum. Such groups include the Symphyla, Collembola and Diplura. In many true insects, especially the more primitive groups such as Thysanura and Blattodea, the flagellum partly or entirely consists of a flexibly connected string of small ring-shaped annuli. The annuli are not true flagellomeres, and in a given insect species the number of annuli generally is not as consistent as the number of flagellomeres in most species.

In many beetles and in the chalcidoid wasps, the apical flagellomeres form a club shape, and the collective term for the segments between the club and the antennal base is the funicle; traditionally in describing beetle anatomy, the term "funicle" refers to the segments between the club and the scape. However, traditionally in working on wasps the funicle is taken to comprise the segments between the club and the pedicel.

Quite commonly the funicle beyond the pedicel is quite complex in Endopterygota such as beetles, moths and Hymenoptera, and one common adaptation is the ability to fold the antenna in the middle, at the joint between the pedicel and the flagellum. This gives an effect like a "knee bend", and such an antenna is said to be geniculate. Geniculate antennae are common in the Coleoptera and Hymenoptera. They are important for insects like ants that follow scent trails, for bees and wasps that need to "sniff" the flowers that they visit, and for beetles such as Scarabaeidae and Curculionidae that need to fold their antennae away when they self-protectively fold up all their limbs in defensive attitudes.

Because the funicle is without intrinsic muscles, it generally must move as a unit, in spite of being articulated. However, some funicles are complex and very mobile. For example, the Scarabaeidae have lamellate antennae that can be folded tightly for safety or spread openly for detecting odours or pheromones. The insect manages such actions by changes in blood pressure, by which it exploits elasticity in walls and membranes in the funicles, which are in effect erectile.

In the groups with more uniform antennae (for example: millipedes), all segments are called *antennomeres*. Some groups have a simple or variously modified apical or subapical bristle called an arista (this may be especially well-developed in various Diptera).

Functions

Olfactory receptors on the antennae bind to free-floating molecules, such as water vapour, and odours including pheromones. The neurons that possess these receptors signal this binding by sending action potentials down their axons to the antennal lobe in the brain. From there, neurons in the antennal lobes connect to mushroom bodies that identify the odour. The sum of the electrical potentials of the antennae to a given odour can be measured using an electroantennogram.

In the monarch butterfly, antennae are necessary for proper time-compensated solar compass orientation during migration. Antennal clocks exist in monarchs, and they are likely to provide the primary timing mechanism for sun compass orientation.

In the African cotton leafworm, antennae have an important function in signaling courtship. Specifically, antennae are required for males to answer the female mating call. Although females do not require antennae for mating, a mating that resulted from a female without antennae was abnormal.

In the diamondback moth, antennae serve to gather information about a host plant's taste and odor. After the desired taste and odor has been identified, the female moth will deposit her eggs onto the plant. Giant swallowtail butterflies also rely on antenna sensitivity to volatile compounds to identify host plants. It was found that females are actually more responsive with their antenna sensing, most likely because they are responsible for oviposition on the correct plant.

In the crepuscular hawk moth (*Manduca sexta*), antennae aid in flight stabilization. Similar to halteres in Dipteran insects, the antennae transmit coriolis forces through the Johnston's organ that can then be used for corrective behavior. A series of low-light, flight stability studies in which moths with flagellae amputated near the pedicel showed significantly decreased flight stability over those with intact antennae. To determine whether there may be other antennal sensory inputs, a second group of moths had their antennae amputated and then re-attached, before being tested in the same stability study. These moths showed slightly decreased performance from intact moths, indicating there are possibly other sensory inputs used in flight stabilization. Re-amputation of the antennae caused a drastic decrease in flight stability to match that of the first amputated group.

Mouthparts

Insects have a range of mouthparts, adapted to particular modes of feeding. The earliest insects had chewing mouthparts. Specialization has mostly been for piercing and sucking, although a range of specializations exist, as these modes of feeding have evolved a number of times (for example, mosquitoes and aphids (which are true bugs) both pierce and suck, however female mosquitoes feed on animal blood whereas aphids feed on plant fluids). In this page, the individual mouthparts are introduced for chewing insects. Specializations are generally described thereafter.

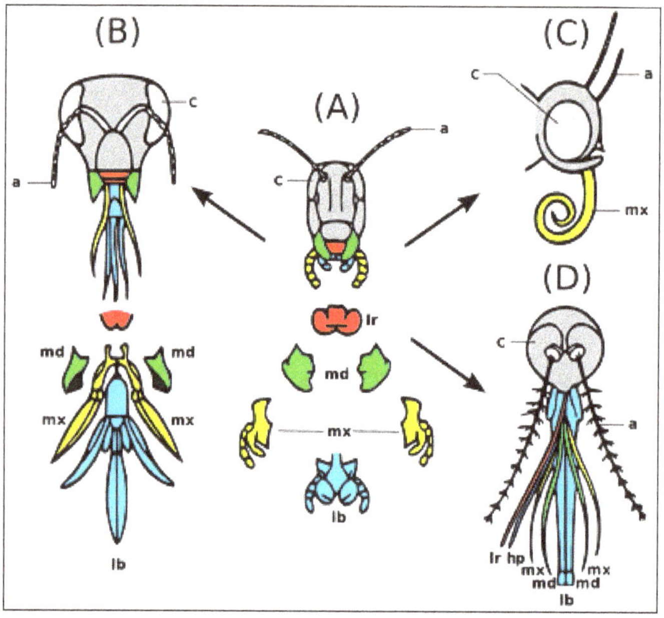

The development of insect mouthparts from the primitive chewing mouthparts of a grasshopper in the centre (A), to the lapping type (B) of a bee, the siphoning type (C) of a butterfly and the sucking type (D) of a female mosquito. Legend: a, antennae; c, compound eye; lb, labium; lr, labrum; md, mandibles; mx, maxillae hp hypopharynx.

Evolution

Like most external features of arthropods, the mouthparts of hexapoda are highly derived. Insect mouthparts show a multitude of different functional mechanisms across the wide diversity of species considered insects. Certainly it is common for significant homology to be conserved, with matching structures formed from matching primordia, and having the same evolutionary origin. On the other hand, even structures that physically are almost identical, and share almost identical functionality as well, may not be homologous; their analogous functions and appearance might be the product of convergent evolution.

Chewing Insects

Examples of chewing insects include dragonflies, grasshoppers and beetles. Some insects do not have chewing mouthparts as adults but do chew solid food when they feed while they still are larvae. The moths and butterflies are major examples of such adaptations.

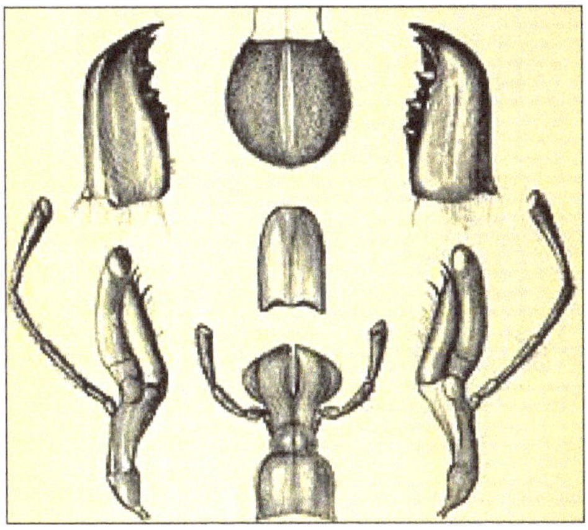

The trophi, or mouthparts of a locust, a typical chewing insect:

- Labrum,
- Mandibles,
- Maxillae,
- Labium,
- Hypopharynx.

Mandible

The mandibles of a bull ant.

European honeybee (Apis mellifera) lapping mouthparts, showing labium and maxillae.

A chewing insect has a pair of mandibles, one on each side of the head. The mandibles are caudal to the labrum and anterior to the maxillae. Typically the mandibles are the largest and most robust mouthparts of a chewing insect, and it uses them to masticate (cut, tear, crush, chew) food items. Two sets of muscles move the mandibles in the coronal plane: abductor muscles move insects' mandibles apart (laterally); adductor muscles bring them together (medially). This they do mainly in opening and closing their jaws in feeding, but also in using the mandibles as tools, or possibly in fighting; note however, that this refers to the coronal plane of the mouth, not necessarily of the insect's body, because insects' heads differ greatly in their orientation.

In carnivorous chewing insects, the mandibles commonly are particularly serrated and knife-like, and often with piercing points. In herbivorous chewing insects mandibles tend to be broader and flatter on their opposing faces, as for example in caterpillars.

In males of some species, such as of Lucanidae and some Cerambycidae, the mandibles are modified to such an extent that they do not serve any feeding function, but are instead used to defend mating sites from other males. In some ants and termites, the mandibles also serve a defensive function (particularly in soldier castes). In bull ants, the mandibles are elongate and toothed, used both as hunting (and defensive) appendages. In bees, which feed primarily by use of a proboscis, the primary use of the mandibles is to manipulate and shape wax, and many paper wasps have mandibles adapted to scraping and ingesting wood fibres.

Maxilla

Situated beneath (caudal to) the mandibles, paired maxillae manipulate and, in chewing insects, partly masticate, food. Each maxilla consists of two parts, the proximal cardo (plural *cardines*), and distal stipes (plural *stipites*). At the apex of each stipes are two lobes, the inner lacinia and outer galea (plurals *laciniae* and *galeae*). At the outer margin, the typical galea is a cupped or scoop-like structure, located over the outer edge of the labium. In non-chewing insects, such as adult Lepidoptera, the maxillae may be drastically adapted to other functions.

Unlike the mandibles, but like the labium, the maxillae bear lateral palps on their stipites. These palps serve as organs of touch and taste in feeding and in the inspection of potential foods and/or prey.

In chewing insects, adductor and abductor muscles extend from inside the cranium to within the bases of the stipites and cardines much as happens with the mandibles in feeding, and also in using the maxillae as tools. To some extent the maxillae are more mobile than the mandibles, and the galeae, laciniae, and palps also can move up and down somewhat, in the sagittal plane, both in feeding and in working, for example in nest building by mud-dauber wasps.

Maxillae in most insects function partly like mandibles in feeding, but they are more mobile and less heavily sclerotised than mandibles, so they are more important in manipulating soft, liquid, or particulate food rather than cutting or crushing food such as material that requires the mandibles to cut or crush.

Like the mandibles, maxillae are innervated by the sub-esopharyngeal ganglia.

Labium

The labium typically is a roughly quadrilateral structure, formed by paired, fused secondary maxillae. It is the major component of the floor of the mouth. Typically, together with the maxillae, the labrum assists manipulation of food during mastication.

Dragonfly nymph feeding on fish that it has caught with its labium and snatched back to the other mouthparts for eating. The labium is just visible from the side, between the front pairs of legs.

The role of the labium in some insects however, is adapted to special functions; perhaps the most dramatic example is in the jaws of the nymphs of the Odonata, the dragonflies and damselflies. In these insects, the labium folds neatly beneath the head and thorax, but the insect can flick it out to snatch prey and bear it back to the head, where the chewing mouthparts can demolish it and swallow the particles.

The labium is attached at the rear end of the structure called cibarium, and its broad basal portion is divided into regions called the submentum, which is the proximal part, the mentum in the middle, and the prementum, which is the distal section, and furthest anterior.

The prementum bears a structure called the ligula; this consists of an inner pair of lobes called glossae and a lateral pair called paraglossae. These structures are homologous to the lacinia and galea of maxillae. The labial palps borne on the sides of labium are the counterparts of maxillary palps. Like the maxillary palps, the labial palps aid sensory function in eating. In many species the musculature of the labium is much more complex than that of the other jaws, because in most, the ligula, palps and prementum all can be moved independently.

The labium is innervated by the sub-esophageal ganglia. In the honey bee, the labium is elongated to form a tube and tongue, and these insects are classified as having both chewing and lapping mouthparts.

The wild silk moth (Bombyx mandarina) is an example of an insect that has small labial palpi and no maxillary palpi.

Hypopharynx

The hypopharynx is a somewhat globular structure, located medially to the mandibles and the maxillae. In many species it is membranous and associated with salivary glands. It assists in swallowing the food. The hypopharynx divides the oral cavity into two parts: the cibarium or dorsal food pouch and ventral salivarium into which the salivary duct opens.

Mandible

The mandibles of a bull ant.

Insect mandibles are a pair of appendages near the insect's mouth, and the most anterior of the three pairs of oral appendages (the labrum is more anterior, but is a single fused structure). Their function is typically to grasp, crush, or cut the insect's food, or to defend against predators or rivals. Insect mandibles, which appear to be evolutionarily derived from legs, move in the horizontal plane unlike those of vertebrates, which appear to be derived from gill arches and move vertically.

Grasshoppers, Crickets and other Simple Insects

The mouthparts of orthopteran insects are often used as a basic example of mandibulate (chewing) mouthparts, and the mandibles themselves are likewise generalized in structure. They are large and hardened, shaped like pinchers, with cutting surfaces on the distal portion and chewing or grinding surfaces basally. They are usually lined with teeth and move sideways. Large pieces of leaves can therefore be cut and then pulverized near the actual mouth opening. This same simple structure is seen in all of the remaining Polyneopteran insect orders, with the exception of the Paraneoptera (Hemiptera, Thysanoptera, and Phthiraptera). Likewise, the mandibles of adult and larval Odonata are simple and generalized, while Ephemeroptera rarely feed as adults, though the larvae ("nymphs") have simple mandibles.

True Bugs

The Hemiptera, and other insects whose mouthparts are described as piercing-sucking, have modified mandibles. Rather than being tooth-like, the mandibles of such insects are lengthened into stylets, which form the outer two parts of the feeding tube, or beak. The mandibles are therefore instrumental in piercing the plant or animal tissues upon which these insects feed, and in helping draw up fluids to the insect's mouth. They are also described as mandibles.

Beetles

Stag beetle with mandibles modified no longer used in feeding.

Nearly all adult beetles, and many beetle larvae, have mandibles. In general form they are similar to those of grasshoppers: Hardened and tooth-like.

Beetle mandibles show a remarkable amount of variability between species, and some are very highly adapted to the food sources or other uses that the species has for them. Certain firefly larvae (family Lampyridae) that feed on snails have grooved mandibles that not only physically break down their prey, but also deliver digestive fluids by these grooves. Ground beetles (family Carabidae) of the tribe Cychrini have long mandibles that project far in front of them, which aid them in feeding on snails inside their shells.

Members of the stag beetle family (Lucanidae) have greatly enlarged mandibles that are often forked, resembling the horns of various deer, from which their common name comes, and similar modifications appear in various scarab beetles and longhorn beetles. Males of these beetles use their mandibles to grasp or displace each other as they compete for mates.

Lice

The mandibles in Phthiraptera (lice) are also modified into piercing stylets.

Thrips

Thysanoptera (thrips) have a variation of piercing mouthparts. In this order only the left mandible is present, modified into a stylet.

Neuropteroids

Within the Neuropterida, adults have chewing mouthparts, but the mandibles of male dobsonflies are non-functional in feeding. The larvae in many lineages are predatory, with mandibles modified with grooves along which digestive saliva flows, while the larvae of the family Sisyridae have the mouthparts developed into a sucking tube which they use to feed on the liquid tissues of freshwater sponges.

Ants, Bees and Wasps

Wasp drinking; the yellow mandibles can be seen on either side of the mouthparts.

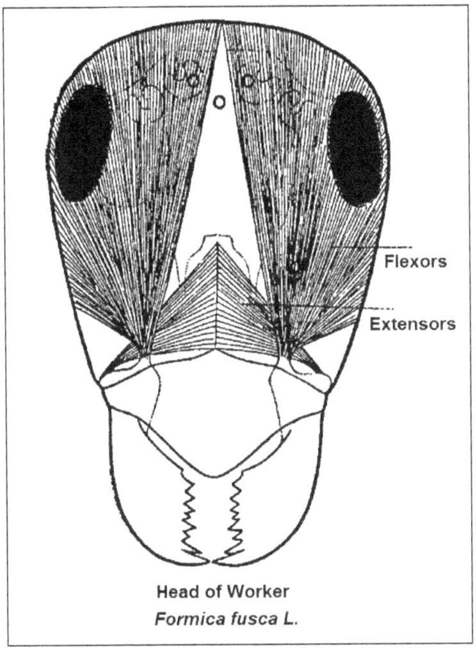

Musculature of ant mandibles.

Most adult Hymenoptera have mandibles that follow the general form, as in grasshoppers. The mandibles are used to clip pieces of vegetation, gather wood fibers, dig nests, or to capture and disassemble prey. What is unusual is that many Hymenoptera have the remaining mouthparts modified to form a proboscis (a "tongue" used to feed on liquids), making them virtually the only insects that normally possess both chewing mouthparts *and* sucking mouthparts (a few exceptional members of other orders may exhibit this, such as flower-feeding beetles that also have "tongues").

Trigona corvina, and other stingless bees, utilize their mandibles for defense purposes and typically interlock them with other individuals while fighting for resources.

Flies

Several families of flies, notably mosquitoes (family Culicidae), have mandibles that are modified into stylets for piercing, similar to the true bugs.

Flies of the Muscomorpha, including the house fly, *Musca domestica*, stable fly, *Stomoxys calcitrans*, blow flies (family Calliphoridae), and many others, lack mandibles altogether, and the mouthparts are designed for sponging up liquids.

Butterflies and Moths

All but a few adult Lepidoptera lack mandibles, with the remaining mouthparts forming an elongated sucking tube. The exception is the mandibulate moths (family Micropterigidae), which have fully developed mandibles as adults.

THORAX

The insect thorax is a surprisingly complex piece of anatomy. All at once, the insect thorax serves as housing for muscles, the mechanism of walking and flight, and, in some cases, serves as the stage on which an insect presents itself, as in the case of the actually harmless Death's Head hawkmoth (Acherontia atropos) seen above with the characteristically skull-shaped pattern on its thorax.

The thorax is the second of three major tagma, or segments that make up an insect's body.

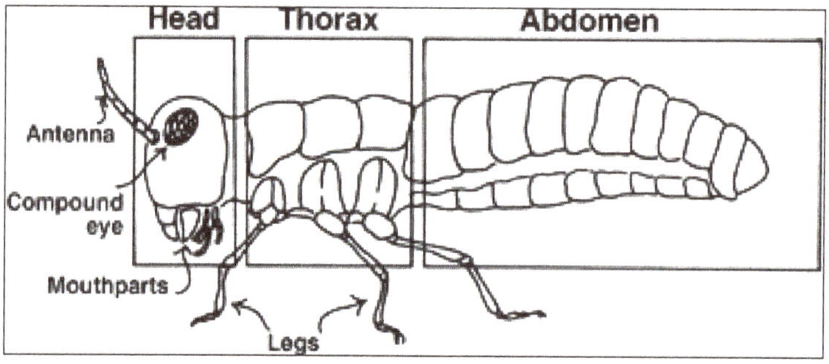

The thorax of an insect is where all appendages for locomotion are, including legs and wings. The thorax consists of three segments, known as the prothorax, the mesothorax, and the metathorax, in that order. Each thoracic segment bears one pair of legs. If the insect is capable of flight, the meso- and metathorax bear the wings; collectively, the wing-bearing segments of insects are called the pteryothoracic segments. The prothorax, which is the very first segment of the thorax, comes after the head, and the cervix, a neck-like membranous region that helps connect muscles from the head to the thorax. The job of the prothorax is usually to serve as a place of attachment for the first pair of legs, although in some insect orders, such as Orthoptera (Grasshoppers), Coleoptera (Beetles), and Blattodea (Cockroaches), the prothorax can also serve as a hard plate, that protects the pteryothoracic segments.

A prothorax that has taken the shape of a hardened plate can be most clearly seen on a grasshopper. The plate covers the cervix, and the anterior part of the mesothorax. The mesothorax comes directly after the prothorax, and bears one pair of legs, and, if the insect is capable of flight, the first pair of wings. The metathorax then comes after the mesothorax, and also bears one pair of legs, and the second pair of wings. The insect thorax has gone through incredible evolution and change in order to accommodate flight. Because not all insects fly, there are differences between the thorax of flying insects, and non-flying insects, as well as differences between the thorax of larval flying insects, and adult flying insects.

Primary Segmentation vs. Secondary Segmentation

Primary segmentation occurs in non-sclerotized insects, or larvae and soft bodied insects. Primary segmentation works by physically shortening the segment in order to move the insect along, like how a caterpillar or inch worm moves.

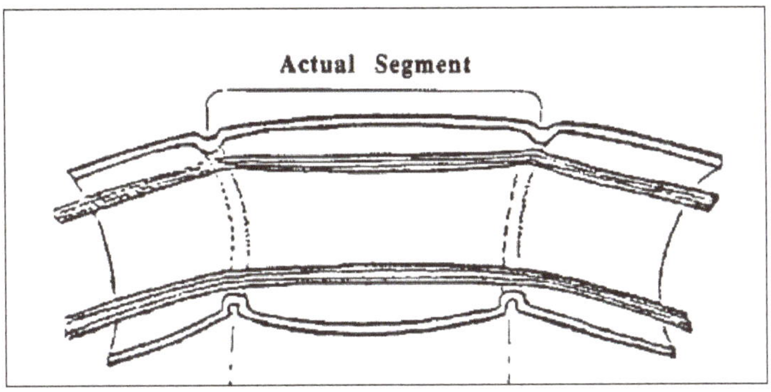

From the diagram on the right, you can see how to muscles within the thoracic are directly attached to the thoracic segment. When those muscles contract, the whole segment contracts, making the inch worm scrunch up.

Secondary segmentation occurs in sclerotized insects, insects that have hard plates over their thoracic segments. The hardened plates most likely evolved as a protective device. Because the hardened plates, or sclerites, cannot move like an inch worm does, a membranous region is created (called the conjuctiva), where the segments telescope into one another. Except for the most primitive lineages, most insects have some degree of secondary segmentation, in the thorax and abdomen. Secondary segmentation is often most evident in the abdomen. Terga overlap each other posteriorly, such that the posterior border of the tergum overlaps with the anterior border of the tergum behind it. Along the anterior edge of each tergum is a costa (a rib or rib-like structure) called the antecosta (and the associated suture is called the antecostal suture, as the name implies, it is a true intersegmental suture). The antecosta, however, is frequently not exactly on the leading edge of the tergum, and the scleritized strip anterior to it is the acrotergite. Immediately anterior to the acrotergite is a membranous region (conjuctiva) marking the region between adjacent terga. From doing comparisons of the terga and sterna structures throughout different insects and arthropods, we see that the antecosta marks the true intersegmental boundary, and the acrotergite belongs to the preceding body terga. The functional dorsal part of the tergum combined with the acrotergite is called the Alinotum.

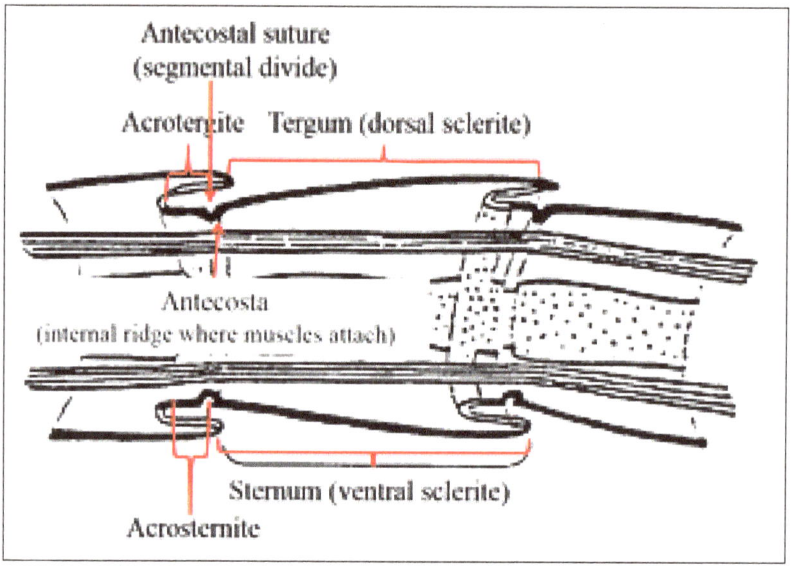

Pterothorax

As I explained above, the pterothorax is the collective name of the meso- and metathorax, which hold the wings. The overall structure of the entire thoracic region is fundamentally the same as the abdomen, but there are modifications for locomotion, and the degree of modification depends on the flying ability of the insect. A very important modification that the thorax has that the abdomen does not is sclerotization in the pleural region.

The pterothroax itself has many extra strengthening ridges, and the conjuctivae is reduced, making the pterothroax a very rigid structure, because the segments cannot telescope. Here is a basic diagram of all the features of the pterothoracic segment. The diagram is being viewed laterally.

Wings

Insect wings are adult outgrowths of the insect exoskeleton that enable insects to fly. They are found on the second and third thoracic segments (the mesothorax and metathorax), and the two pairs are often referred to as the forewings and hindwings, respectively, though a few insects lack hindwings, even rudiments. The wings are strengthened by a number of longitudinal veins, which often have cross-connections that form closed "cells" in the membrane (extreme examples include the dragonflies and lacewings). The patterns resulting from the fusion and cross-connection of the wing veins are often diagnostic for different evolutionary lineages and can be used for identification to the family or even genus level in many orders of insects.

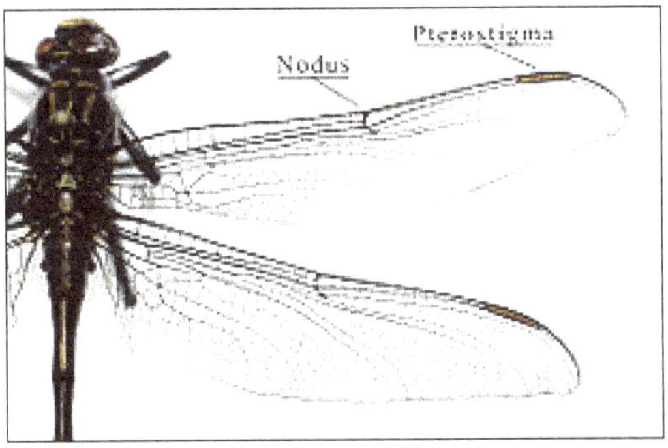

Original veins and wing posture of a dragonfly.

Hoverflies hovering to mate.

Physically, some insects move their flight muscles directly, others indirectly. In insects with direct flight, the wing muscles directly attach to the wing base, so that a small downward movement of the wing base lifts the wing itself upward. Those insects with indirect flight have muscles that attach to and deform the thorax, causing the wings to move as well.

The wings are present in only one sex (often the male) in some groups such as velvet ants and Strepsiptera, or are selectively lost in "workers" of social insects such as ants and termites. Rarely, the female is winged but the male not, as in fig wasps. In some cases, wings are produced only at particular times in the life cycle, such as in the dispersal phase of aphids. Wing structure and colouration often vary with morphs, such as in the aphids, migratory phases of locusts and polymorphic butterflies. At rest, the wings may be held flat, or folded a number of times along specific patterns; most typically, it is the hindwings which are folded, but in a few groups such as the vespid wasps, it is the forewings.

How and why insect wings evolved is not well understood and there has been a long standing debate about their origins. During the 19th century, the question of insect wing evolution originally rested on two main positions. One position postulated insect wings evolved from pre-existing structures, while the second proposed insect wings were entirely novel formations. The "novel" hypothesis suggested that insect wings did not form from pre-existing ancestral appendages but rather as outgrowths from the insect body wall.

A cockchafer's hardened forewings raised, hindwings unfolding.

Bumblebee's wing.

Long since, research on insect wing origins has built on the "pre-existing structures" position that was originally proposed in the 19th century. Recent literature has pointed to several ancestral structures as being important to the origin of insect wings. Among these include: gills, respiratory appendages of legs, and lateral (paranotal) and postero-lateral projections of the thorax to name a few. According to more current literature, gill-like structures and the paranotal lobe still appear to be among the most important ancestral structures to insect wing origins.

Today, there are three main theories on the origins of insect flight. These theories are referred to as the paranotal lobe theory, the gill theory and the dual theory of insect wing evolution. These theories postulate that wings either developed from paranotal lobes, extensions of the thoracic terga; that they are modifications of movable abdominal gills as found on aquatic naiads of mayflies; or that insect wings arose from the fusion of pre-existing endite and exite structures each with pre-existing articulation and tracheation.

Morphology

Internal

Each of the wings consists of a thin membrane supported by a system of veins. The membrane is formed by two layers of integument closely apposed, while the veins are formed where the two layers remain separate; sometimes the lower cuticle is thicker and more heavily sclerotized under a vein. Within each of the major veins there is a nerve and a trachea, and, since the cavities of the veins are connected with the hemocoel, hemolymph can flow into the wings.

As the wing develops, the dorsal and ventral integumental layers become closely apposed over most of their area forming the wing membrane. The remaining areas form channels, the future veins, in which the nerves and tracheae may occur. The cuticle surrounding the veins becomes thickened and more heavily sclerotized to provide strength and rigidity to the wing. Two types of hair may occur on the wings: microtrichia, which are small and irregularly scattered, and macrotrichia, which are larger, socketed, and may be restricted to veins. The scales of Lepidoptera and Trichoptera are highly modified macrotrichia.

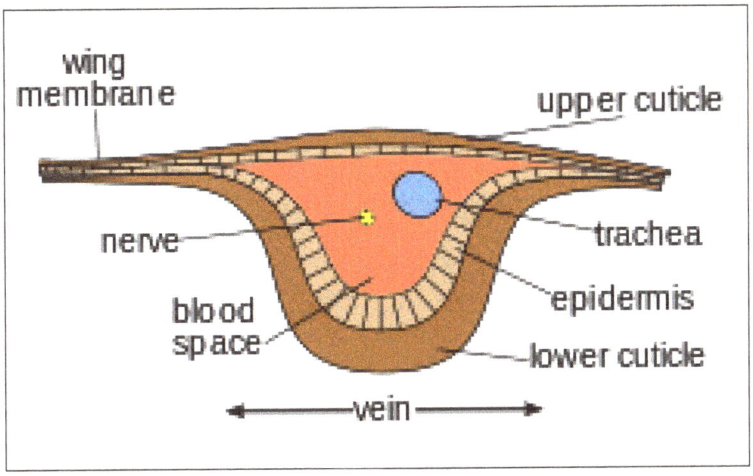

Venation

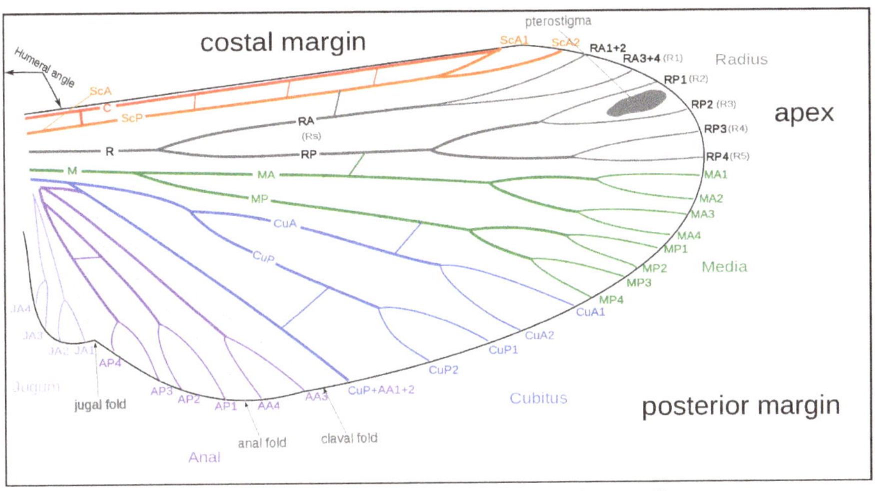

Venation of insect wings, based on the Comstock–Needham system.

In some very small insects, the venation may be greatly reduced. In Chalcidoidea (Chalcid wasps), for instance, only the subcosta and part of the radius are present. Conversely, an increase in venation may occur by the branching of existing veins to produce accessory veins or by the development of additional, intercalary veins between the original ones, as in the wings of Orthoptera (grasshoppers and crickets). Large numbers of cross-veins are present in some insects, and they may form a reticulum as in the wings of Odonata (dragonflies and damselflies) and at the base of the forewings of Tettigonioidea and Acridoidea (katydids and grasshoppers respectively).

The archedictyon is the name given to a hypothetical scheme of wing venation proposed for the very first winged insect. It is based on a combination of speculation and fossil data. Since all winged insects are believed to have evolved from a common ancestor, the archedictyon represents the "template" that has been modified (and streamlined) by natural selection for 200 million years. According to current dogma, the archedictyon contained 6–8 longitudinal veins. These veins (and their branches) are named according to a system devised by John Comstock and George Needham—the Comstock–Needham system:

- Costa (C): The leading edge of the wing.
- Subcosta (Sc): Second longitudinal vein (behind the costa), typically unbranched.
- Radius (R): Third longitudinal vein, one to five branches reach the wing margin.
- Media (M): Fourth longitudinal vein, one to four branches reach the wing margin.
- Cubitus (Cu) :Fifth longitudinal vein, one to three branches reach the wing margin.
- Anal veins (A1, A2, A3: Unbranched veins behind the cubitus.

The costa (C) is the leading marginal vein on most insects, although sometimes there is a small vein above the costa called the precosta, although in almost all extant insects, the precosta is fused with the costa. The costa rarely ever branches because it is at the leading edge, which is associated at its base with the humeral plate. The trachea of the costal vein is perhaps a branch of the subcostal trachea. Located after the costa is the third vein, the subcosta, which branches into two separate veins: The anterior and posterior. The base of the subcosta is associated with the distal end of the neck of the first axillary. The fourth vein is the radius (R), which is branched into five separate veins. The radius is generally the strongest vein of the wing. Toward the middle of the wing, it forks into a first undivided branch (R1) and a second branch, called the radial sector (Ra), which subdivides dichotomously into four distal branches (R2, R3, R4, R5). Basally, the radius is flexibly united with the anterior end of the second axillary (2Ax).

The fifth vein of the wing is the media. In the archetype pattern (A), the media forks into two main branches: a media anterior (MA), which divides into two distal branches (MA1, MA2), and a median sector, or media posterior (MP), which has four terminal branches (M1, M2, M3, M4). In most modern insects the media anterior has been lost, and the usual "media" is the four-branched media posterior with the common basal stem. In the Ephemerida, according to present interpretations of the wing venation, both branches of the media are retained, while in Odonata the persisting media is the primitive anterior branch. The stem of the media is often united with the radius, but when it occurs as a distinct vein its base is associated with the distal median plate (m') or is continuously sclerotized with the latter. The cubitus, the sixth vein of the wing, is primarily two branched. The primary forking of the takes place near the base of the wing, forming the two principal branches (Cu1, Cu2). The anterior branch may break up into a number of secondary branches, but commonly it forks into two distal branches. The second branch of the cubitus (Cu2) in Hymenoptera, Trichoptera, and Lepidoptera was mistaken by Comstock and Needham for the first anal. Proximally the main stem of the cubitus is associated with the distal median plate (m') of the wing base.

Postcubitus (Pcu) is the first anal of the Comstock–Needham system. The postcubitus, however, has the status of an independent wing vein and should be recognized as such. In nymphal wings, its trachea arises between the cubital trachea and the group of vannal tracheae. In the mature wings of more generalized insect the Postcubitus is always associated proximally with the cubitus and is never intimately connected with the flexor sclerite (3Ax) of the wing base. In Neuroptera, Mecoptera, and Trichoptera the postcubitus may be more closely associated with the vannal veins, but its base is always free from the latter. The postcubitus is usually unbranched; it is primitively two branched. The vannal veins (lV to nV) are the anal veins that are immediately associated with the third axillary, and which are directly affected by the movement of this sclerite that brings about the flexion of the wings. In number the vannal veins vary. from 1 to 12, according to the expansion of the vannal area of the wing. The vannal tracheae usually arise from a common tracheal stem in nymphal insects, and the veins are regarded as branches of a single anal vein. Distally the vannal veins are either simple or branched. Jugal Veins (J) of the jugal lobe of the wing is often occupied by a network of irregular veins, or it may be entirely membranous; but sometimes it contains one or two distinct small veins, the first jugal vein, or vena arcuata, and the second jugal vein, or vena cardinalis (2J).

- C-Sc cross-veins: Run between the costa and subcostal.
- R cross-veins: Run between adjacent branches of the radius.
- R-M cross-veins: Run between the radius and media.
- M-Cu cross-veins: Run between the media and cubitus.

All the veins of the wing are subject to secondary forking and to union by cross-veins. In some orders of insects the cross-veins are so numerous that the whole venational pattern becomes a close network of branching veins and cross-veins. Ordinarily, however, there is a definite number of cross-veins having specific locations. The more constant cross-veins are the humeral cross-vein (h) between costa and subcosta, the radial cross-vein (r) between R and the first fork of Rs, the sectorial cross-vein (s) between the two forks of R8, the median cross-vein (m–m) between M2 and M3, and the mediocubital cross-vein (m-cu) between media and cubitus.

The veins of insect wings are characterized by a convex-concave placement, such as those seen in mayflies (i.e., concave is "down" and convex is "up") which alternate regularly and by its triadic type of branching; whenever a vein forks there is always an interpolated vein of the opposite position between the two branches. A concave vein will fork into two concave veins (with the interpolated vein being convex) and the regular alteration of the veins is preserved. The veins of the wing appear to fall into an undulating pattern according to whether they have a tendency to fold up or down when the wing is relaxed. The basal shafts of the veins are convex, but each vein forks distally into an anterior convex branch and a posterior concave branch. Thus the costa and subcosta are regarded as convex and concave branches of a primary first vein, Rs is the concave branch of the radius, posterior media the concave branch of the media, Cu1 and Cu2 are respectively convex and concave, while the primitive Postcubitus and the first vannal have each an anterior convex branch and a posterior concave branch. The convex or concave nature of the veins has been used as evidence in determining the identities of the persisting distal branches of the veins of modern insects, but it has not been demonstrated to be consistent for all wings.

Fields

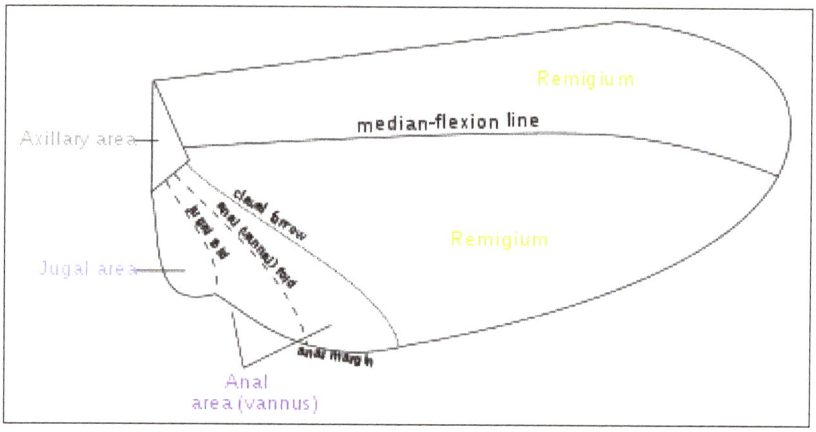

Wing areas are delimited and subdivided by fold-lines along which the wing can fold, and flexion-lines along which the wing can flex during flight. The fundamental distinction between the flexion-lines and the fold-lines is often blurred, as fold-lines may permit some flexibility or vice versa. Two constants that are found in nearly all insect wings are the claval (a flexion-line) and jugal folds (or fold line); forming variable and unsatisfactory boundaries. Wing foldings can very complicated, with transverse folding occurs in the hindwings of Dermaptera and Coleoptera, and in some insects the anal area can be folded like a fan. There are about four different fields found on the insect wings:

- Remigium,
- Anal area (vannus),

- Jugal area,
- Axillary area,
- Alula.

Most veins and crossveins occur in the anterior area of the remigium, which is responsible for most of the flight, powered by the thoracic muscles. The posterior portion of the remigium is sometimes called the clavus; the two other posterior fields are the anal and jugal ares. When the vannal fold has the usual position anterior to the group of anal veins, the remigium contains the costal, subcostal, radial, medial, cubital, and postcubital veins. In the flexed wing the remigiumturns posteriorly on the flexible basal connection of the radius with the second axillary, and the base of the mediocubital field is folded medially on the axillary region along the plica basalis (bf) between the median plates (m, m') of the wing base.

The vannus is bordered by the vannal fold, which typically occurs between the postcubitus and the first vannal vein. In Orthoptera it usually has this position. In the forewing of Blattidae, however, the only fold in this part of the wing lies immediately before the postcubitus. In Plecoptera the vannal fold is posterior to the postcubitus, but proximally it crosses the base of the first vannal vein. In the cicada the vannal fold lies immediately behind the first vannal vein (lV). These small variations in the actual position of the vannal fold, however, do not affect the unity of action of the vannal veins, controlled by the flexor sclerite (3Ax), in the flexion of the wing. In the hindwings of most Orthoptera a secondary vena dividens forms a rib in the vannal fold. The vannus is usually triangular in shape, and its veins typically spread out from the third axillary like the ribs of a fan. Some of the vannal veins may be branched, and secondary veins may alternate with the primary veins. The vannal region is usually best developed in the hindwing, in which it may be enlarged to form a sustaining surface, as in Plecoptera and Orthoptera. The great fanlike expansions of the hindwings of Acrididae are clearly the vannal regions, since their veins are all supported on the third axillary sclerites on the wing bases, though Martynov ascribes most of the fan areas in Acrididae to the jugal regions of the wings. The true jugum of the acridid wing is represented only by the small membrane (Ju) mesad of the last vannal vein. The jugum is more highly developed in some other Polyneoptera, as in the Mantidae. In most of the higher insects with narrow wings the vannus becomes reduced, and the vannal fold is lost, but even in such cases the flexed wing may bend along a line between the postcubitus and the first vannal vein.

The Jugal Region, or Neala, is a region of the wing that is usually a small membranous area proximal to the base of the vannus strengthened by a few small, irregular veinlike thickenings; but when well developed it is a distinct section of the wing and may contain one or two jugal veins. When the jugal area of the forewing is developed as a free lobe, it projects beneath the humeral angle of the hindwing and thus serves to yoke the two wings together. In the Jugatae group of Lepidoptera it bears a long finger-like lobe. The jugal region was termed the neala ("new wing") because it is evidently a secondary and recently developed part of the wing.

The axillary region is region containing the axillary sclerites has in general the form of a scalene triangle. The base of the triangle (a-b) is the hinge of the wing with the body; the apex (c) is the distal end of the third axillary sclerite; the longer side is anterior to the apex. The point d on the anterior side of the triangle marks the articulation of the radial vein with the second axillary sclerite. The line between d and c is the plica basalis (bf), or fold of the wing at the base of the mediocubital field.

At the posterior angle of the wing base in some Diptera there is a pair of membranous lobes (squamae, or calypteres) known as the alula. The alula is well developed in the house fly. The outer squama (c) arises from the wing base behind the third axillary sclerite (3Ax) and evidently represents the jugal lobe of other insects (A, D); the larger inner squama (d) arises from the posterior scutellar margin of the tergum of the wing-bearing segment and forms a protective, hoodlike canopy over the halter. In the flexed wing the outer squama of the alula is turned upside down above the inner squama, the latter not being affected by the movement of the wing. In many Diptera a deep incision of the anal area of the wing membrane behind the single vannal vein sets off a proximal alar lobe distal to the outer squama of the alula.

Joints

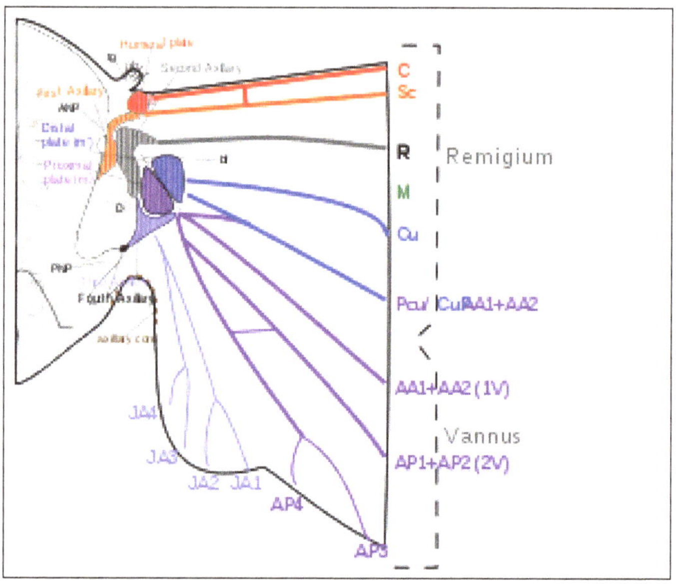

The various movements of the wings, especially in insects that flex the wings horizontally over the back when at rest, demand a more complicated articular structure at the wing base than a mere hinge of the wing with the body. Each wing is attached to the body by a membranous basal area, but the articular membrane contains a number of small articular sclerites, collectively known as the pteralia. The pteralia include an anterior humeral plate at the base of the costal vein, a group of axillaries (Ax) associated with the subcostal, radial, and vannal veins, and two less definite median plates (m, m') at the base of the mediocubital area. The axillaries are specifically developed only in the wing-flexing insects, where they constitute the flexor mechanism of the wing operated by the flexor muscle arising on the pleuron. Characteristic of the wing base is also a small lobe on the anterior margin of the articular area proximal to the humeral plate, which, in the forewing of some insects, is developed into a large, flat, scale-like flap, the tegula, overlapping the base of the wing. Posteriorly the articular membrane often forms an ample lobe between the wing and the body, and its margin is generally thickened and corrugated, giving the appearance of a ligament, the so-called axillary cord, continuous mesally with the posterior marginal scutellar fold of the tergal plate bearing the wing.

The articular sclerites, or pteralia, of the wing base of the wing-flexing insects and their relations to the body and the wing veins, shown diagrammatically, are as follows:

- Humeral plates,
- First Axillary,
- Second Axillary,
- Third Axillary,
- Fourth Axillary,
- Median plates (m, m').

The humeral plate is usually a small sclerite on the anterior margin of the wing base, movable and articulated with the base of the costal vein. Odonata have their humeral plate greatly enlarged, with two muscles arising from the episternum inserted into the Humeral plates and two from the edge of the epimeron inserted into the axillary plate.

The first axillary sclerite (lAx) is the anterior hinge plate of the wing base. Its anterior part is supported on the anterior notal wing process of the tergum (ANP); its posterior part articulates with the tergal margin. The anterior end of the sclerite is generally produced as a slender arm, the apex of which (e) is always associated with the base of the subcostal vein (Sc), though it is not united with the latter. The body of the sclerite articulates laterally with the second axillary. The second axillary sclerite (2Ax) is more variable in form than the first axillary, but its mechanical relations are no less definite. It is obliquely hinged to the outer margin of the body of the first axillary, and the radial vein (R) is always flexibly attached to its anterior end (d). The second axillary presents both a dorsal and a ventral sclerotization in the wing base; its ventral surface rests upon the fulcral wing process of the pleuron. The second axillary, therefore, is the pivotal sclerite of the wing base, and it specifically manipulates the radial vein.

The third axillary sclerite (3Ax) lies in the posterior part of the articular region of the wing. Its form is highly variable and often irregular, but the third axillary is the sclerite on which is inserted the flexor muscle of the wing (D). Mesally it articulates anteriorly (f) with the posterior end of the second axillary, and posteriorly (b) with the posterior wing process of the tergum (PNP), or with a small fourth axillary when the latter is present. Distally the third axillary is prolonged in a process which is always associated with the bases of the group of veins in the anal region of the wing here termed the vannal veins (V). The third axillary, therefore, is usually the posterior hinge plate of the wing base and is the active sclerite of the flexor mechanism, which directly manipulates the vannal veins. The contraction of the flexor muscle (D) revolves the third axillary on its mesal articulations (b, f) and thereby lifts its distal arm; this movement produces the flexion of the wing. The Fourth Axillary sclerite is not a constant element of the wing base. When present it is usually a small plate intervening between the third axillary and the posterior notal wing process and is probably a detached piece of the latter.

The median plates (m, m') are also sclerites that are not so definitely differentiated as specific plates as are the three principal axillaries, but nevertheless they are important elements of the flexor apparatus. They lie in the median area of the wing base distal to the second and third axillaries and are separated from each other by an oblique line (bf) which forms a prominent convex fold during flexion of the wing. The proximal plate (m) is usually attached to the distal arm of the third axillary and perhaps should be regarded as a part of the latter. The distal plate (m') is less constantly present as a distinct sclerite and may be represented by a general sclerotization of the base of the mediocubital field of the wing. When the veins of this region are distinct at their bases, they are associated with the outer median plate.

Muscles

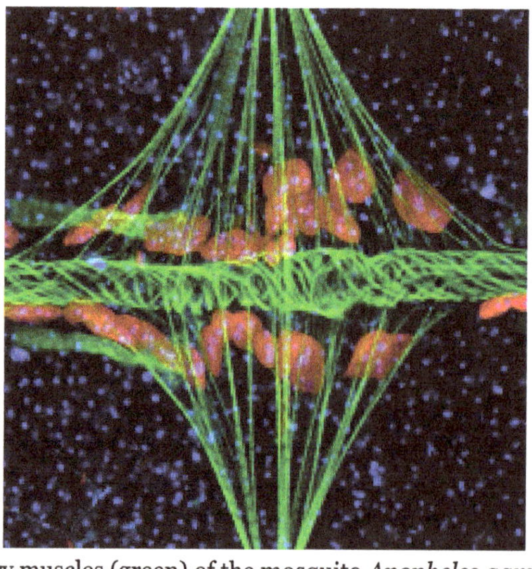

The diamond-shaped alary muscles (green) of the mosquito *Anopheles gambiae* and their structural relationship to the tube-like heart (also in green). Red depicts pericardial cells, blue cell nuclei.

The muscles that control flight in insects can take up to 10% to 30% of the total body mass. The muscles that control flight vary with the two types of flight found in insects: indirect and direct. Insects that use first, indirect, have the muscles attach to the tergum instead of the wings, as the name suggests. As the muscles contract, the thoracic box becomes distorted, transferring the energy to the wing. There are two "bundles" of muscles, those that span parallel to the tergum, the dorsolongitudinals, and those that are attached to the tegum and extend to the sternum, the dorsoventrals. In direct muscle, the connection is directly from the pleuron (thoracic wall) to individual sclerites located at the base of the wing. The subalar and basalar muscles have ligament attachments to the subalar and basalar sclerites. Here resilin, a highly elastic material, forms the ligaments connecting flight muscles to the wing apparatus.

In more derived orders of insects, such as Diptera (flies) and Hymenoptera (wasp), the indirect muscles occupy the greatest volume of the pterothorax and function as the primary source of power for the wingstroke. Contraction of the dorsolongitudinal muscles causes the severe arching of the notum which depresses the wing while contraction of the dorsoventral muscles causes opposite motion of notum. Other more primitive insects, such as Orthoptera (locusts), Coleoptera (beetles), and Odonata (dragonflies) use direct muscles that are responsible for developing the needed power for the up and down strokes.

Insect wing muscle is a strictly aerobic tissue. Per unit protein it consumes fuel and oxygen at rates taking place in a very concentrated and highly organized tissue so that the steady-state rates per unit volume represent an absolute record in biology. The fuel and oxygen rich blood is carried to the muscles through diffusion occurring in large amounts, in order to maintain the high level of energy used during flight. Many wing muscles are large and may be as large as 10 mm in length and 2 mm in width. Moreover, in some Diptera the fibres are of giant dimensions. For instance, in the very active *Rutilia*, the cross-section is 1800 µm long and more than 500 µm wide. The transport of fuel and oxygen from the surroundings to the sites of consumption and the reverse transport of carbon dioxide therefore represent a challenge to the biologist both in relation to transport in the liquid phase and in the intricate system of air tubes, i.e. in the tracheal system.

Coupling, Folding and other Features

In many insect species, the forewing and hindwing are coupled together, which improves the aerodynamic efficiency of flight. The most common coupling mechanism (e.g., Hymenoptera and Trichoptera) is a row of small hooks on the forward margin of the hindwing, or "hamuli", which lock onto the forewing, keeping them held together (hamulate coupling). In some other insect species (e.g., Mecoptera, Lepidoptera, and some Trichoptera) the jugal lobe of the forewing covers a portion of the hindwing (jugal coupling), or the margins of the forewing and hindwing overlap broadly (amplexiform coupling), or the hindwing bristles, or frenulum, hook under the retaining structure or retinaculum on the forewing.

When at rest, the wings are held over the back in most insects, which may involve longitudinal folding of the wing membrane and sometimes also transverse folding. Folding may sometimes occur along the flexion lines. Though fold lines may be transverse, as in the hindwings of beetles and earwigs, they are normally radial to the base of the wing, allowing adjacent sections of a wing to be folded over or under each other. The commonest fold line is the jugal fold, situated just behind the third anal vein, although, most Neoptera have a jugal fold just behind vein 3A on the forewings. It is sometimes also present on the hindwings. Where the anal area of the hindwing is large, as in Orthoptera and Blattodea, the whole of this part may be folded under the anterior part of the wing along a vannal fold a little posterior to the claval furrow. In addition, in Orthoptera and Blattodea, the anal area is folded like a fan along the veins, the anal veins being convex, at the crests of the folds, and the accessory veins concave. Whereas the claval furrow and jugal fold are probably homologous in different species, the vannal fold varies in position in different taxa. Folding is produced by a muscle arising on the pleuron and inserted into the third axillary sclerite in such a way that, when it contracts, the sclerite pivots about its points of articulation with the posterior notal process and the second axillary sclerite.

As a result, the distal arm of the third axillary sclerite rotates upwards and inwards, so that finally its position is completely reversed. The anal veins are articulated with this sclerite in such a way that when it moves they are carried with it and become flexed over the back of the insect. Activity of the same muscle in flight affects the power output of the wing and so it is also important in flight control. In orthopteroid insects, the elasticity of the cuticle causes the vannal area of the wing to fold along the veins. Consequently, energy is expended in unfolding this region when the wings are moved to the flight position. In general, wing extension probably results from the contraction of muscles attached to the basalar sclerite or, in some insects, to the subalar sclerite.

Flight

Flight Mechanisms

Australian emperor in flight; it uses the direct flight mechanism.

Two groups of relatively large insects, the Ephemeroptera (mayflies) and the Odonata (dragonflies and damselflies) have the flight muscles attached directly to their wings;

the wings can beat no faster than the rate at which nerves can send impulses to command the muscles to beat. All other living winged insects fly using a different mechanism, involving indirect flight muscles which cause the thorax to vibrate; the wings can beat faster than the rate at which the muscles receive nerve impulses. This mechanism evolved once, and is the defining feature (synapomorphy) for the infraclass Neoptera.

Aerodynamics

There are two basic aerodynamic models of insect flight. Most insects use a method that creates a spiralling leading edge vortex. Some very small insects use the fling and clap or Weis-Fogh mechanism in which the wings clap together above the insect's body and then fling apart. As they fling open, the air gets sucked in and creates a vortex over each wing. This bound vortex then moves across the wing and, in the clap, acts as the starting vortex for the other wing. Circulation and lift are increased, at the price of wear and tear on the wings.

Many insects can hover by beating their wings rapidly, requiring sideways stabilization as well as lift.

A few insects use gliding flight, without the use of thrust. It is found in some species of arboreal ants, known as gliding ants.

Evolution

Sometime in the Carboniferous Period, some 350 million years ago, when there were only two major land masses, insects began flying. How and why insect wings developed, however, is not well understood, largely due to the scarcity of appropriate fossils from the period of their development in the Lower Carboniferous. Three main theories on the origins of insect flight are that wings developed from paranotal lobes, extensions of the thoracic terga; that they are modifications of movable abdominal gills as found on aquatic naiads of mayflies; or that they developed from thoracic protrusions used as radiators.

Fossils

Fossils from the Devonian (400 million years ago) are all wingless, but by the Carboniferous (320 million years ago), more than 10 different genera of insects had fully functional wings. There is little preservation of transitional forms between the two periods. The earliest winged insects are from this time period (Pterygota), including the Blattoptera, Caloneurodea, primitive stem-group Ephemeropterans, Orthoptera and Palaeodictyopteroidea. Very early Blattopterans (during the Carboniferous) had a very large discoid pronotum and coriaceous forewings with a distinct CuP vein (an unbranched wing vein, lying near the claval fold and reaching the wing posterior margin). Even though the oldest definitive insect fossil is the Devonian *Rhyniognatha hirsti*, estimated at 396–407 million years old, it possessed dicondylic mandibles, a feature associated with winged insects.

Holotype wing of the extinct *Cimbrophlebia brooksi.* 49.5 Million Years old; "Boot Hill", Klondike Mountain Formation, Washington, USA.

During the Permian, the dragonflies (Odonata) were the dominant aerial predator and probably dominated terrestrial insect predation as well. True Odonata appeared in the Permian and all are amphibious. Their prototypes are the oldest winged fossils, go back to the Devonian, and are different from other wings in every way. Their prototypes may have had the beginnings of many modern attributes even by late Carboniferous and it is possible that they even captured small vertebrates, for some species had a wing span of 71 cm. The earliest beetle-like species during the Permian had pointed, leather like forewings with cells and pits. Hemiptera, or true bugs had appeared in the form of Arctiniscytina and Paraknightia having forewings with unusual venation, possibly diverging from Blattoptera.

A single large wing from a species of Diptera in the Triassic (10 mm instead of usual 2–6 mm) was found in Australia (Mt. Crosby). This family Tilliardipteridae, despite of the numerous 'tipuloid' features, should be included in Psychodomorpha sensu Hennig on account of loss of the convex distal 1A reaching wing margin and formation of the anal loop.

Morphogenesis

While the development of wings in insects is clearly defined in those who are members of Endopterygota, which undergo complete metamorphosis; in these species, the wing develops while in the pupal stage of the insects life cycle. However, insects that undergo incomplete metamorphosis do not have a pupal stage, therefore they must have a different wing morphogenesis. Insects such as those that are hemimetabolic have wings that start out as buds, which are found underneath the exoskeleton, and do not become exposed until the last instar of the nymph.

The first indication of the wing buds is of a thickening of the hypodermis, which can be observed in insect species as early the embryo, and in the earliest stages of the life cycle. During the development of morphological features while in the embryo, or embryogenesis, a cluster of cells grow underneath the ectoderm which later in development, after the lateral ectoderm has grown dorsally to form wind imaginal disc. An example of wing bud development in the larvae, can be seen in those of White butterflies (*Pieris*). In the second instar the histoblast become more prominent, which now form a pocket-like structure. As of the third and fourth instars, the histoblast become more elongated. This greatly extended and evaginated, or protruding, part is what becomes the wing. By the close of the last instar, or fifth, the wing is pushed out of the wing-pocket, although continues to lie under the old larval cuticle while in its prepupal stage. It is not until the butterfly is in its pupal stage that the wing-bud becomes exposed, and shortly after eclosion, the wing begins to expand and form its definitive shape.

The development of tracheation of the wings begin before the wing histoblast form, as it is important to note that they develop near a large trachea. During the fourth instar, cells from the epithelium of this trachea become greatly enlarged extend into the cavity of the wing bud, with each cell having developed a closely coiled tracheole. Each trachcole is of unicellular origin, and is at first intracellular in position; while tracheae are of multicellular origin and the lumen of each is intercellular in position. The development of tracheoles, each coiled within a single cell of the epithelium of a trachea, and the subsequent opening of communication between the tracheoles and the lumen of the trachea, and the uncoiling and stretching out of the tracheoles, so that they reach all parts of the wing.

In the earlier stages of its development, the wing-bud is not provided with special organs of respiration such as tracheation, as it resembles in this respect the other portions of the hypodermis of which it is still a part. It should be noted, however, that the histoblast is developed near a large trachea, a cross-section of which is shown in, which represents sections of these parts of the first, second, third and fourth instars respectively. At the same time the tracheoles uncoil, and extend in bundles in the forming vein-cavities of the wing-bud. At the molt that marks the beginning of the pupal stadium stage, they become functional. At the same time, the larval tracheoles degenerate; their function having been replaced by the wing tracheae.

Adaptations

Variation

Insect wings are fundamental in identifying and classifying species as there is no other set of structures in studying insects more significant. Each order and insect family has distinctive wing shapes and features. In many cases, even species may be distinguished from each other by differences of color and pattern. For example, just by position one can identify species, albeit to a much lesser extent. Though most insects fold their wings when at rest, dragonflies and some damselflies rest with their wings spread out horizontally, while groups such as the caddisflies, stoneflies, alderflies, and lacewings hold their wings sloped roof-like over their backs. A few moths wrap their wings around their bodies, while many flies and most butterflies close their wings together straight upward over the back.

Many times the shape of the wings correlates with the type of insect flight. The best-flying insects tend to have long, slender wings. In many species of Sphingidae (sphinx moths), the forewings are large and sharply pointed, forming with the small hindwings a triangle that is suggestive of the wings of fast, modern airplanes. Another, possibly more important correlation, is that of the size and power of the muscles to the speed and power of flight. In the powerfully flying insects, the wings are most adapted for the stresses and aerodynamics of flight. The veins are thicker, stronger, and closer together toward the front edge (or "leading edge") and thinner yet flexible toward the rear edge (or "trailing edge"). This makes the insect wing an excellently constructed airfoil, capable of exerting both propulsion and lift while minimizing drag.

Variation of the wing beat may also occur, not just amongst different species, but even among individuals at different times. In general, the frequency is dependent upon the ratio between the power of the wing muscles and the resistance of the load. Large-winged, light-bodied butterflies may have a wing beat frequency of 4–20 per second whereas small-winged, heavy-bodied flies and bees beat their wings more than 100 times a second and mosquitoes can beat up to 988–1046 times a second. The same goes for flight; though it is generally difficult to estimate the speed of insects in flight, most insects can probably fly faster in nature than they do in controlled experiments.

Coleoptera

In species of Coleoptera (beetles), the only functional wings are the hindwings. The hindwings are longer than the elytra, folded longitudinally and transversely under the elytra. The wing is rotated forwards on its base into flight position. This action spread the wing and unfolded longitudinally and transversely. There is the spring mechanism in the wing structure, sometimes with the help of abdomen movement, to keep the wing in folded position. The beetle wing venation is reduced and modified due to the folding structure, which include:

Cross folding in the wings of beetles	
	The hindwing, spread: by folding lines, it is divided into five fields that are completed each to the rear.

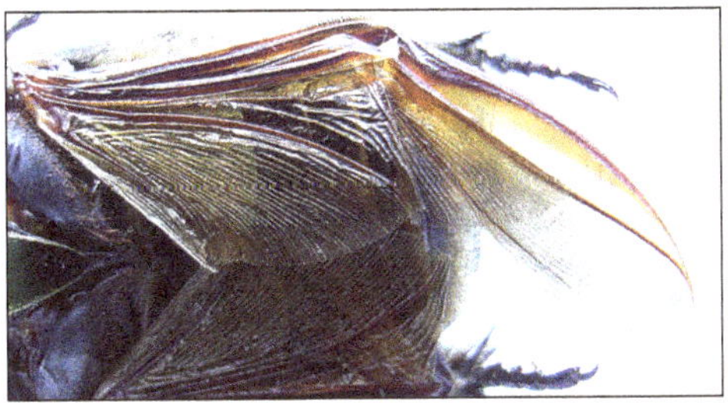	The same wing, half folded: The two joints of the cross-folding form an obtuse angle. The right is already in the wings folded in three layers. With greater resolution, the third arch of the wing margin in the first and second is visible. To the left of the fifth arch appears in the fourth.
	The same wing, folded completely: The five fields are aligned (The elytra have been removed).

- Costa (C), Subcosta posterior (ScP): At the leading wing marginal, fused for most of the length.
- Radius anterior (RA): Divided into two branches beyond the middle of the wing.
- Radius posterior (RP): Basal connection is lost.
- Media posterior (MP): Branches, long and strong vein.
- Cubitus anterior (CuA).
- Anal veins (AA, AP): Veins behind the cubitus, separated by anal fold.

In most species of beetles, the front pair of wings are modified and sclerotised (hardened) to form elytra and they protect the delicate hindwings which are folded beneath. The elytra are connected to the pterathorax; being called as such because it is where the wings are connected. The elytra are not used for flight, but tend to cover the hind part of the body and protect the second pair of wings (*alae*). The elytra must be raised in order to move the hind flight wings. A beetle's flight wings are crossed with veins and are folded after landing, often along these veins, and are stored below the elytra. In some beetles, the ability to fly has been lost. These include some ground beetles (family Carabidae) and some "true weevils" (family Curculionidae), but also some desert and cave-dwelling species of other families. Many of these species have the two elytra fused together, forming a solid shield over the abdomen. In a few families, both the ability to fly and the elytra have been lost, with the best known example being the glow-worms of the family Phengodidae, in which the females are larviform throughout their lives.

Lepidoptera

The two pairs of wings are found on the middle and third segment, or mesothorax and metathorax respectively. In the more recent genera, the wings of the second segment are much more pronounced, however some more primitive forms have similarly sized wings of both segments. The wings are covered in scales arranged like shingles, forming the extraordinary variety seen in color. The mesothorax is evolved to have more powerful muscles to propel moth or butterfly through the air, with the wing of said segment having a stronger vein structure. The largest superfamily, Noctuidae, has the wings modified to act as tympanal or hearing organs Modifications in the wing's venation include:

- Costa (C): Not found in Butterflies.
- Subcosta (Sc) + Radius 1 (Sc+R1): At the leading wing marginal, fused or very close for most of the length, in hindwing fused and well developed in the humeral area, subcosta never branches in butterfly.
- Radius (R2-R5): Radius divides into branches beyond the middle of the wing up to five branches in Papilionidae. On forewing, the last R is stalked in all butterflies except Hesperiidae is separated.
- Radius sector (Rs): In hindwing.
- Media (M1-M3): The basal section has been lost.
- Cubitus anterior (CuA1-CuA2): CuP section has been lost.
- Anal veins (A, 1A+2A, 3A): Either one vein A, or two veins 1A+2A, 3A.
- Humeral vein: The hindwing of most butterflies has the humeral vein, except Lycaenidae There is the enlargement of the humeral area of the hindwing which is overlapped with the forewing. The humeral vein strengthened the hindwing overlapped area so that the two wings coupling better.

The wings, head parts of thorax and abdomen of Lepidoptera are covered with minute scales, from which feature the order 'Lepidoptera' derives its names. Most scales are lamellar, or blade-like and attached with a pedicel, while other forms may be hair-like or specialized as secondary sexual characteristics. The lumen or surface of the lamella, has a complex structure. It gives color either due to the pigmentary colors contained within or due to its three-dimensional structure. Scales provide a number of functions, which include insulation, thermoregulation, aiding gliding flight, amongst others, the most important of which is the large diversity of vivid or indistinct patterns they provide which help the organism protect itself by camouflage, mimicry, and to seek mates.

Odonata

Species of Odonata (Damselflies and dragonflies) both have two pairs of wings which are about equal in size and shape and are clear in color. There are five, if the R+M is counted as 1, main vein stems on dragonfly and damselfly wings, and wing veins are fused at their bases and the wings cannot be folded over the body at rest, which also include:

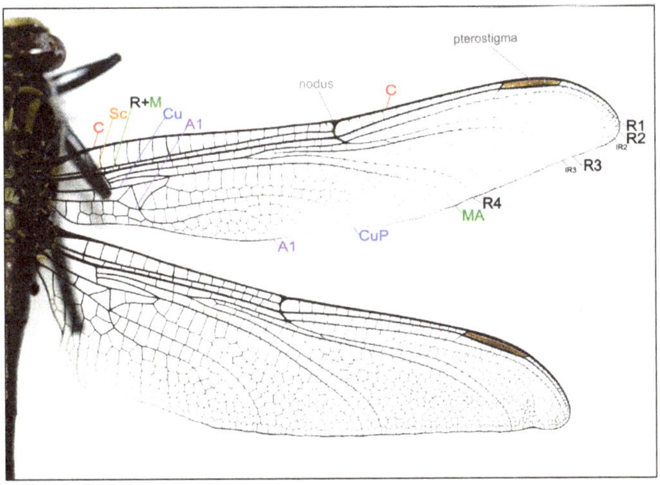

Dragonfly.

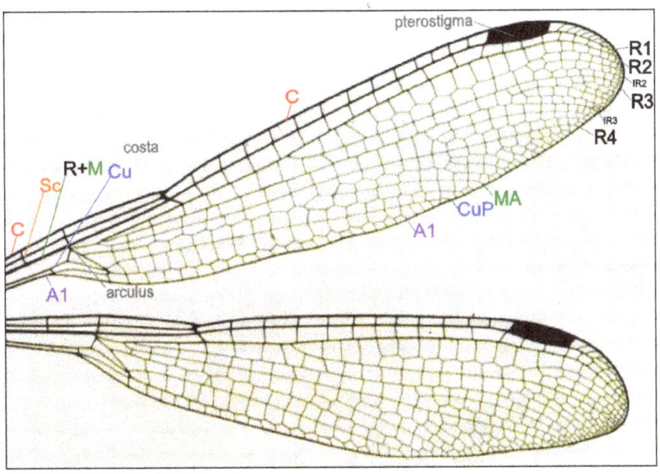

Damselfly.

- Costa (C): At the leading edge of the wing, strong and marginal, extends to the apex of the wing.
- Subcosta (Sc): Second longitudinal vein, it is unbranched, joins C at nodus.
- Radius and Media (R+M): Third and fourth longitudinal vein, the strongest vein on the wing, with branches, R1-R4, reach the wing margin, the media anterior (MA) are also reach the wing margin. IR2 and IR3 are intercalary veins behind R2 and R3 respectively.
- Cubitus (Cu): Fifth longitudinal vein, cubitus posterior (CuP) is unbranched and reach the wing margin.

- Anal veins (A1): Unbranched veins behind the cubitus.
- A nodus is formed where the second main vein meets the leading edge of the wing. The black pterostigma is carried near the wing tip.

The main veins and the crossveins form the wing venation pattern. The venation patterns are different in different species. There may be very numerous crossveins or rather few. The Australian Flatwing Damselfly's wings are one of the few veins patterns. The venation pattern is useful for species identification. Almost all Anisoptera settle with the wings held out sideways or slightly downward, however most Zygoptera settle with the wings held together, dorsal surfaces apposed. The thorax of Zygoptera is so oblique that when held in this way the wings fit neatly along the top of the abdomen. They do not appear to be held straight up as in butterflies or mayflies. In a few zygopteran families the wings are held horizontally at rest, and in one anisopteran genus (e.g. Cordulephya, Corduliidae) the wings are held in the typical damselfly resting position. Adult species possess two pairs of equal or subequal wings. There appear to be only five main vein stems. A nodus is formed where the second main vein (subcosta) meets the leading edge of the wing. In most families a conspicuous pterostigma is carried near the wing tip. Identification as Odonata can be based on the venation. The only likely confusion is with some lacewings (order Neuroptera) which have many crossveins in the wings. Until the early years of the 20th century Odonata were often regarded as being related to lacewings and were given the ordinal name Paraneuroptera, but any resemblance between these two orders is entirely superficial. In Anisoptera the hindwing is broader than the forewing and in both wings a crossvein divides the discoidal cell into a Triangle and Supertriangle.

Orthoptera

Species of Orthoptera (grasshoppers and crickets) have forewings that are tough opaque tegmina, narrow which are normally covering the hindwings and abdomen at rest. The hindwings are board membranous and folded in fan-like manner, which include the following venation:

- Costa (C): At the leading marginal of the forewing and hindwing, unbranched.
- Subcosta (Sc): Second longitudinal vein, unbranched.
- Radius (R): Third longitudinal vein, branched to Rs in forewing and hindwing.
- Media anterior (MA): Fourth longitudinal vein, branched in basal part as Media posterior (MP).
- Cubitus (Cu): Fifth longitudinal vein, on forewing and hindwing dividing near the wing base into branched CuA, and unbranched CuP.
- Anal veins (A): Veins behind the cubitus, unbranched, two in forewing, many in hindwing.

Phasmatodea

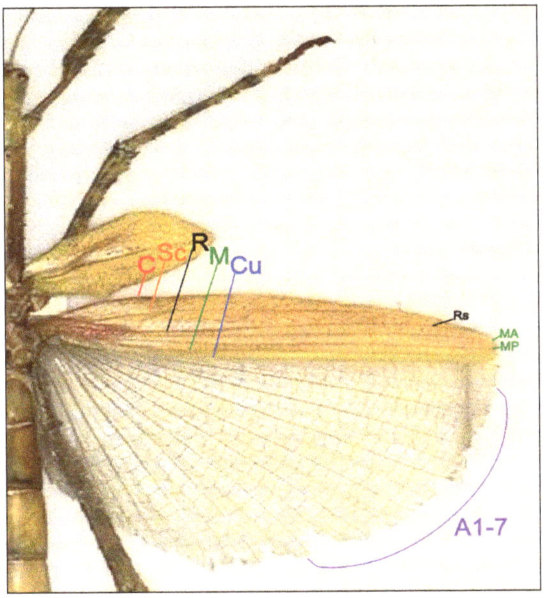

- Costa (C): At the leading marginal of the hindwing, unbranched, absent in forewing.
- Subcosta (Sc): Second longitudinal vein, unbranched.
- Radius (R): Third longitudinal vein, branched to Rs in hindwing, unbranched in forewing.
- Media anterior (MA): Fourth longitudinal vein, branched in basal part as Media posterior (MP).

- Cubitus (Cu): Fifth longitudinal vein, unbranched.
- Anal veins (A): Veins behind the cubitus, unbranched, two in forewing, many in hindwing 1A-7A in one group and the rest in another group.

Stick insect have forewings that are tough, opaque tegmina, short and covering only the base part of the hindwings at rest. Hindwings from costa to Cubitus are tough and opaque like the forewings. The large anal area are membranous and folded in fan-like manner. There are no or very few branching in stick insect wing veins.

Dermaptera

Unfolding of earwig wing	
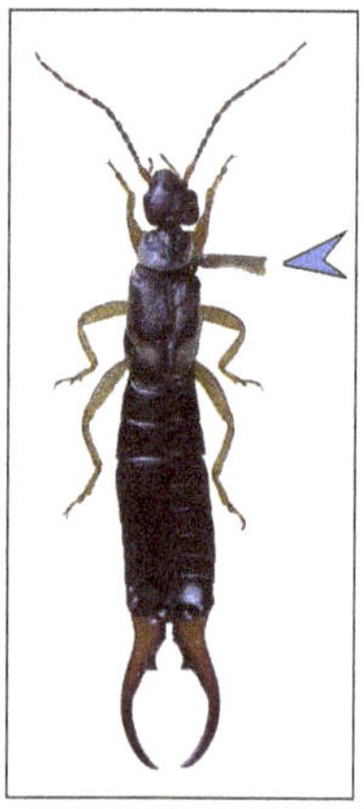	The front and rear wings at rest: The front wing covers most of the hindwing, with only the joint projects in the form of a quarter circle forward with a central white spot under the forewing. On the right hand side of the forewing is opened to the right (blue arrow), which from this perspective appears narrower than it is with the rear wing still folded completely..
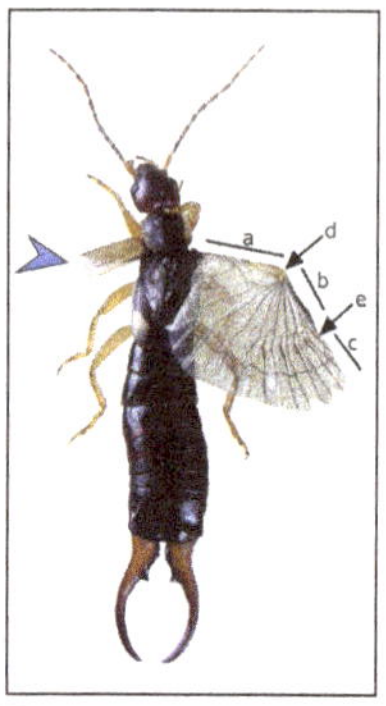	The front wing is open to the left (blue arrow) with the right side of the forewing removed; the hindwing is half open. With greater resolution, the multiple folding is shown, resembling a fan which is parallel to the lines b and c. The arrow points to the e point where the fan is closed again, having been folded by 180°.

Other orders such as the Dermaptera (earwigs), Orthoptera (grasshoppers, crickets), Mantodea (praying mantis) and Blattodea (cockroaches) have rigid leathery forewings that aren't flapped while flying, sometimes called *tegmen* (pl. *tegmina*), *elytra*, or *pseudoelytron*.

Hemiptera

In Hemiptera (true bugs), the forewings may be hardened, though to a lesser extent than in the beetles. For example, the anterior part of the front wings of stink bugs is hardened, while the posterior part is membranous. They are called *hemelytron* (pl. *hemelytra*). They are only found in the suborder Heteroptera; the wings of the Homoptera, such as the cicada, are typically entirely membranous. Both forewings and hindwings of Cicada are membranous. Most species are glass-like although some are opaque. Cicadas are not good fliers and most fly only a few seconds. When flying, forewing and hindwing are hooked together by a grooved coupling along the hindwing costa and forewing margin. Most species have a basic venation as shown in the following picture.

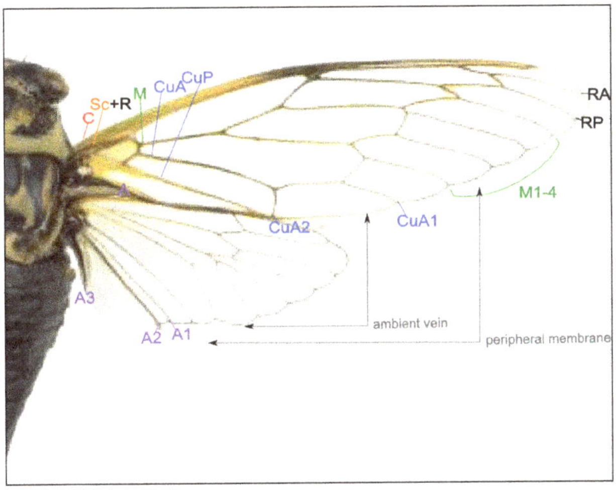

- Costa (C): At the leading wing marginal, in forewing extends to the node and lies close to Sc+R.
- Subcosta + Radius (Sc+R): In forewing Sc and R fused together to the node. Radial sector (Rs) arises near the node and unbranches.
- Radius anterior (RA).
- Radius posterior (RP).
- Media (M): Branches to M1 to M4.
- Cubitus anterior (CuA): Branches to CuA1 and CuA2.
- Cubitus posterior (CuP): Unbranches.
- Anal veins (A): Veins behind the cubitus, 1A and 2A fused in the forewing, CuP and 2A are folded.

Also notice there are the ambient veins and peripheral membranes on the margin of both wings.

Diptera

In the Diptera (true flies), there is only one pair of functional wings, with the posterior pair of wings are reduced to halteres, which help the fly to sense its orientation and movement, as well as to improve balance by acting similar to gyroscopes. In Calyptratae, the very hindmost portion of the wings are modified into somewhat thickened flaps called calypters which cover the halteres.

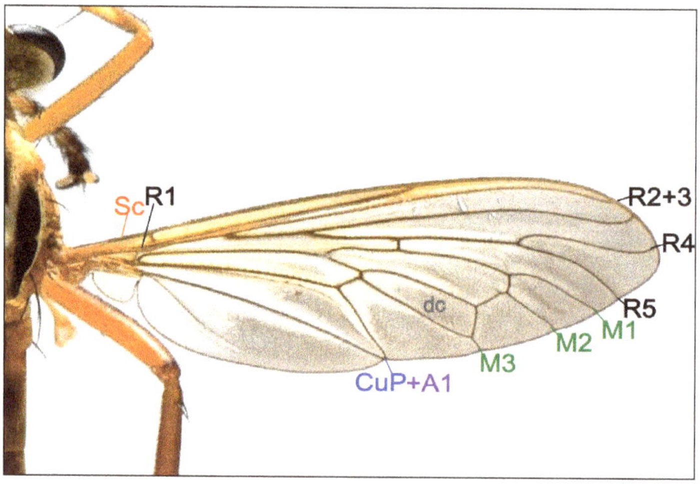

- Costa (C): Not found in Diptera.
- Subcosta (Sc): Became the leading wing vein, unbranched.
- Radius (R): Branched to R1-R5.
- Media (M): Branched to M1-M4.
- Cubitus anterior(CuA): Unbranched, CuP is reduced in Diptera. Some species CuA and 1A are separated, some species meets when reaching the wing margin, some species fused.
- Anal veins (A): Only two anal veins 1A and 2A are present, 2A is not distinctive in some species.
- Discal Cell (dc): Well defined in most species.

Blattodea

Species of Blattodea (cockroaches) have a forewing, are also known as tegmen, that is more or less sclerotized. It is used in flight as well as a form of protection of the membranous hindwings. The veins of hindwing are about the same as front wing but with large anal lobe folded at rest between CuP and 1A. The anal lobe usually folded in a fan-like manner.

- Costa (C): At the leading edge of the wing.
- Subcosta (Sc): Second longitudinal vein, it is relatively short.

- Radius (R): Third longitudinal vein, with many pectinate branches.
- Media (M): Fourth longitudinal vein, reach the wing margin.
- Cubitus anterior (CuA): Fifth longitudinal vein, with dichotomous branches occupy large part of tegmen.

- Cubitus posterior (CuP): Is unbranched, curved and reach the wing margin.
- Anal veins (A): Veins behind the cubitus.

Hymenoptera

The Hymenoptera adults, include sawflies, wasps, bees and non-working ants, all of which have two pairs of membranous wings.

An example of Longitudinal folding in wasps (Vespidae)	
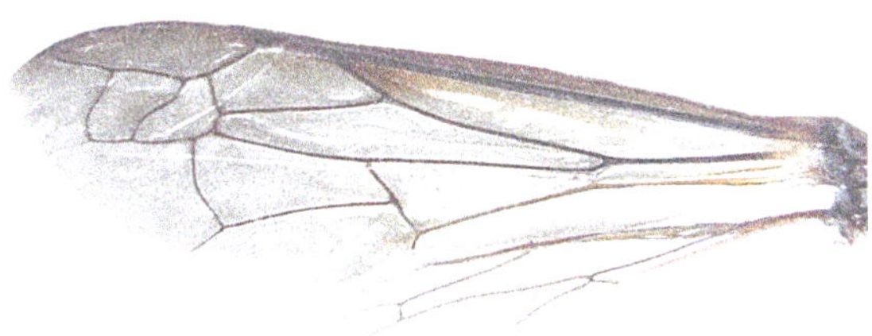	The main fold line of the forewing seen half way up as a bright horizontal line. The wing part that is behind this line is turned back down. The narrow strip at the front edge of the wing is in front of the first strong wire folded forward and down.
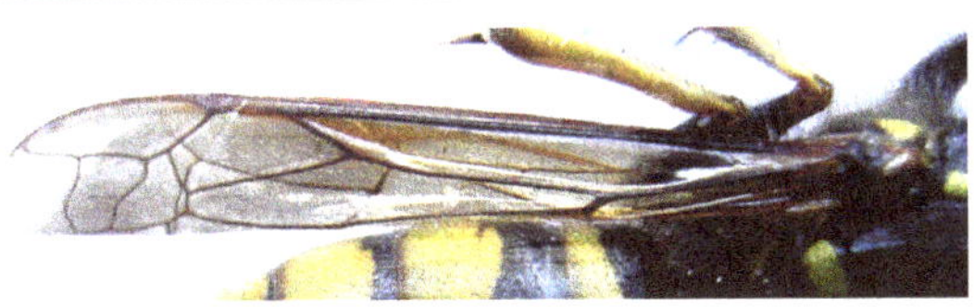	So in rest position, the outer lining forms the tough outer edge of the wing, which protects the sides of the abdomen as a shock absorber. The rear wing is covered largely by the forewing.

- Costa (C): Not found in Hymenoptera.
- Subcosta (Sc): Unbranched.
- Radius (R): Branched to R1-R5.
- Media (M): M is unbranched, in forewing M is fused with Rs for part of its length.
- Cubitus (CuA): Unbranched, CuP is absent in Hymenoptera.
- Anal veins (A): Only two anal veins 1A and 2A are present, 2A is not distinctive in some species.
- Wing-coupling: Row of hooks on the leading edge of hindwing engage the hind margin of the forewing, strongly couple the wings in flight.
- Line of wing folding: Some species, including Vespidae, the forewing are longitudinally folded along the 'line of wing folding' at rest.
- Pterostigma: It is present for some species.

The forward margin of the hindwing bears a number of hooked bristles, or "hamuli", which lock onto the forewing, keeping them held together. The smaller species may have only two or three hamuli on each side, but the largest wasps may have a considerable number, keeping the wings gripped together especially tightly. Hymenopteran wings have relatively few veins compared with many other insects, especially in the smaller species.

Other Families

Termites are relatively poor fliers and are readily blown downwind in wind speeds of less than 2 km/h, shedding their wings soon after landing at an acceptable site, where they mate and attempt to form a nest in damp timber or earth. Wings of most termites have three heavy veins along the basal part of the front edge of the forewing and the crossveins near the wing tip are angled, making trapezoidal cells. Although subterranean termite wings have just two major veins along the front edge of the forewing and the cross veins towards the wingtip are perpendicular to these veins, making square and rectangular cells.

Species of Thysanoptera (thrips) have slender front and hindwings with long fringes of hair, called fringed wings. While species of Trichoptera (caddisfly) have hairy wings with the front and hindwings clothed with setae.

Legs

The arthropod leg is a form of jointed appendage of arthropods, usually used for walking. Many of the terms used for arthropod leg segments (called podomeres) and may be confused with terms for bones: *coxa, trochanter, femur, tibia, tarsus, ischium, metatarsus, carpus, dactylus, patella.*

Homologies of leg segments between groups are difficult to prove and are the source of much argument. Some authors posit up to eleven segments per leg for the most recent common ancestor of extant arthropods but modern arthropods have eight or fewer. It has been argued that the ancestral leg need not have been so complex, and that other events, such as successive loss of function of a *Hox*-gene, could result in parallel gains of leg segments.

Biramous and Uniramous

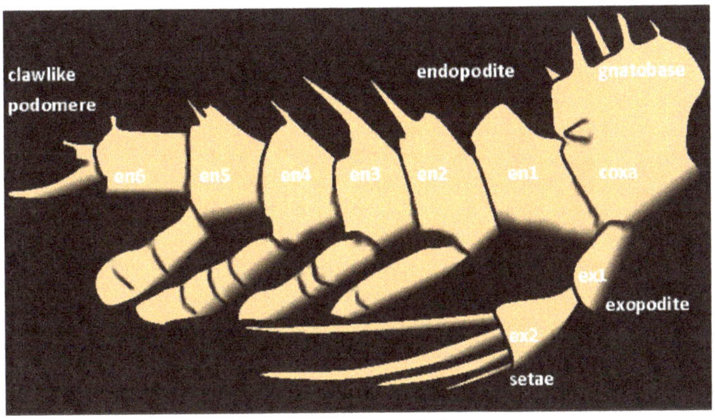

Diagram of biramous leg of a trilobite; *Agnostus* spp.

The appendages of arthropods may be either biramous or uniramous. A uniramous limb comprises a single series of segments attached end-to-end. A biramous limb, however, branches into two, and each branch consists of a series of segments attached end-to-end.

The external branch (ramus) of the appendages of crustaceans is known as the exopod or exopodite, while the internal branch is known as the endopod or endopodite. Other structures aside from the latter two are termed exites (outer structures) and endites (inner structures). Exopodites can be easily distinguished from exites by the possession of internal musculature. The exopodites can sometimes be missing in some crustacean groups (amphipods and isopods), and they are completely absent in insects.

The legs of insects and myriapods are uniramous. In crustaceans, the first antennae are uniramous, but the second antennae are biramous, as are the legs in most species.

For a time, possession of uniramous limbs was believed to be a shared, derived character, so uniramous arthropods were grouped into a taxon called Uniramia. It is now believed that several groups of arthropods evolved uniramous limbs independently from ancestors with biramous limbs, so this taxon is no longer used.

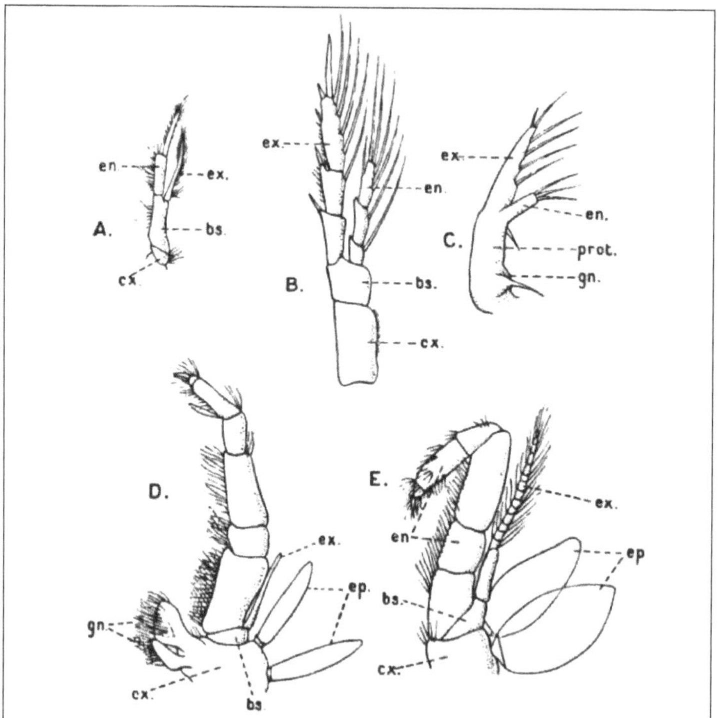

Various types of Crustacean limbs. A, abdominal limb (pleopod) of Crayfish. (After Huxley.) B, swimming-foot of *Calanus* (Copepoda). (After Sars.) C, limb of second pair (antenna) of nauplius larva of *Apus* (Branchiopoda). (After Claus.) D, first thoracic limb of *Anaspides* (Syncarida). E, second thoracic limb of *Anaspides*. *bs*, basipodite ; *cx*, coxopodite ; *en*, endopodite ; *ep*, epipodite ; *ex*, exopodite ; *gn*, gnathobase ; *prot*, protopodite.

Crustacean appendages.

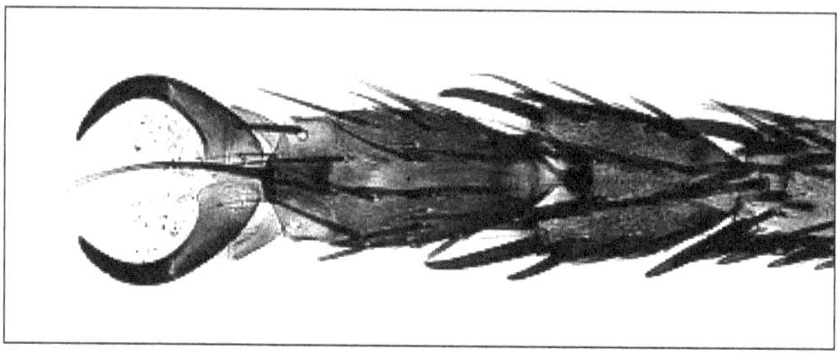

Micrograph of Housefly leg.

Chelicerata

Arachnid legs differ from those of insects by the addition of two segments on either side of the tibia, the patella between the femur and the tibia, and the metatarsus (sometimes called basitarsus) between the tibia and the tarsus (sometimes called telotarsus), making a total of seven segments.

The situation is identical in scorpions, but with the addition of a pre-tarsus beyond the tarsus. The claws of the scorpion are not truly legs, but are pedipalps, a different kind of appendage that is also found in spiders and is specialised for predation and mating.

In *Limulus*, there are no metatarsi or pretarsi, leaving six segments per leg.

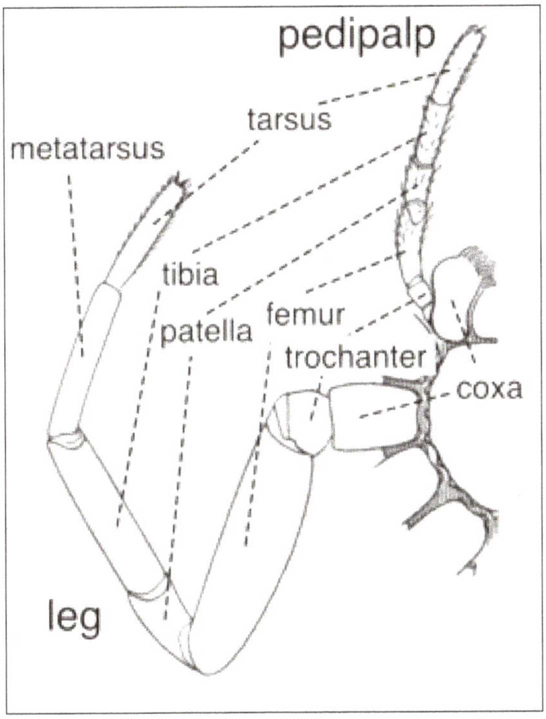

Diagram of a spider leg and pedipalp – the pedipalp has one fewer segment.

Crustacea

The legs of crustaceans are divided primitively into seven segments, which do not follow the naming system used in the other groups. They are: Coxa, basis, ischium, merus, carpus, propodus, and dactylus. In some groups, some of the limb segments may be fused together. The claw (chela) of a lobster or crab is formed by the articulation of the dactylus against an outgrowth of the propodus. Crustacean limbs also differ in being biramous, whereas all other extant arthropods have uniramous limbs.

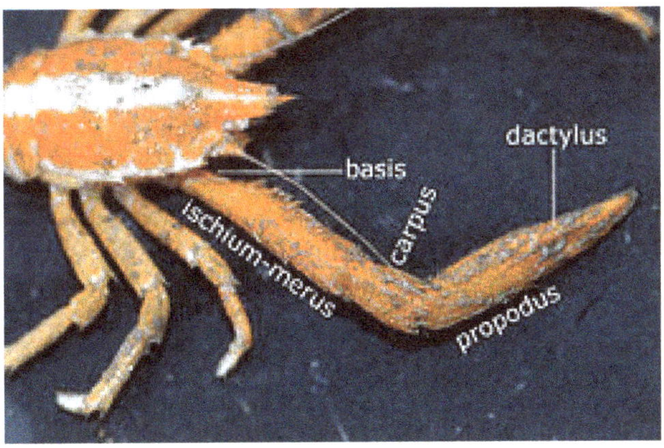

The leg of a squat lobster, showing the segments; the ischium and merus are fused in many decapods.

Myriapoda

Myriapods (millipedes, centipedes and their relatives) have seven-segmented walking legs, comprising coxa, trochanter, prefemur, femur, tibia, tarsus, and a tarsal claw. Myriapod legs show a variety of modifications in different groups. In all centipedes, the first pair of legs is modified into a pair of venomous fangs called forcipules. In most millipedes, one or two pairs of walking legs in adult males are modified into sperm-transferring structures called gonopods. In some millipedes, the first leg pair in males may be reduced to tiny hooks or stubs, while in others the first pair may be enlarged.

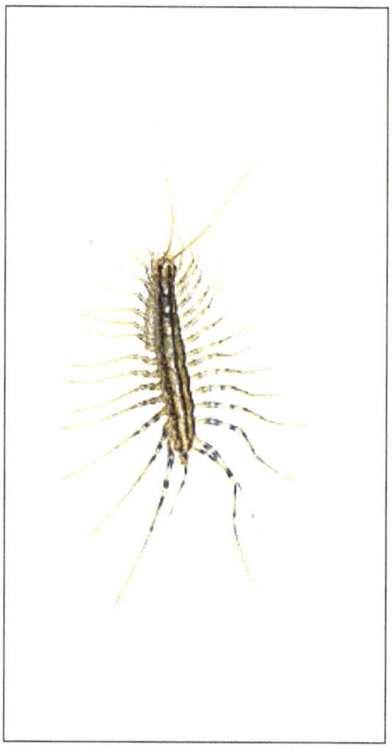

Seven-segmented legs of Scutigera coleoptrata.

Insects

Insects and their relatives are hexapods, having six legs, connected to the thorax, each with five components. In order from the body they are the coxa, trochanter, femur, tibia, and tarsus. Each is a single segment, except the tarsus which can be from three to seven segments, each referred to as a *tarsomere*.

Fundamental Morphology of Insect Legs

A representative insect leg, such as that of a housefly or cockroach, has the following parts, in sequence from most proximal to most distal:

- Coxa,
- Trochanter,
- Femur,

- Tibia,
- Tarsus,
- Pretarsus.

Associated with the leg itself there are various sclerites around its base. Their functions are articular and have to do with how the leg attaches to the main exoskeleton of the insect. Such sclerites differ considerably between unrelated insects.

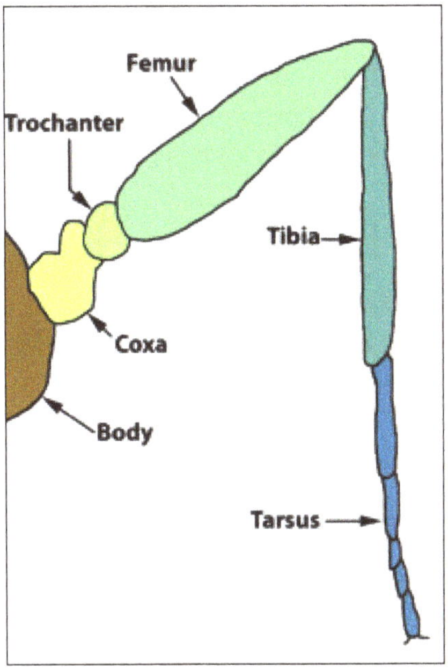

Diagram of a typical insect leg.

Coxa

Zabalius aridus showing full leg anatomy, including plantulae under each tarsomere.

The coxa is the proximal segment and functional base of the leg. It articulates with the pleuron and associated sclerites of its thoracic segment, and in some species it articulates with the edge of the sternite as well. The homologies of the various basal sclerites are open to debate. Some authorities suggest that they derive from an ancestral subcoxa. In many species the coxa has two lobes where it articulates with the pleuron. The posterior lobe is the meron which is usually the larger part of the coxa. A meron is well developed in Periplaneta, the Isoptera, Neuroptera and Lepidoptera.

Trochanter

The trochanter articulates with the coxa but usually is attached rigidly to the femur. In some insects its appearance may be confusing; for example it has two subsegments in the Odonata. In parasitic Hymenoptera the base of the femur has the appearance of a second trochanter.

Femur

In most insects the femur is the largest region of the leg; it is especially conspicuous in many insects with saltatorial legs because the typical leaping mechanism is to straighten the joint between the femur and the tibia, and the femur contains the necessary massive bipennate musculature.

Acanthacris ruficornis, legs saltatorial, femora with bipennate muscle attachments, spines on tibiae painfully effective in a defensive kick.

Tibia

The tibia is the fourth section of the typical insect leg. As a rule the tibia of an insect is slender in comparison to the femur, but it generally is at least as long and often longer. Near the distal end there is generally a tibial spur, often two or more. In the Apocrita the tibia of the foreleg bears a large apical spur that fits over a semicircular gap in the first segment of the tarsus. The gap is lined with comb-like bristles, and the insect cleans its antennae by drawing them through.

Tarsus

Robber fly (Asilidae), showing tarsomeres and pretarsi with ungues, pulvilli and empodia

The ancestral tarsus was a single segment and in the extant Protura, Diplura and certain insect larvae the tarsus also is single-segmented. Most modern insects have tarsi divided into subsegments (tarsomeres), usually about five. The actual number varies with the taxon, which may be useful for diagnostic purposes. For example, the Pterogeniidae characteristically have 5-segmented fore- and mid-tarsi, but 4-segmented hind tarsi, whereas the Cerylonidae have four tarsomeres on each tarsus.

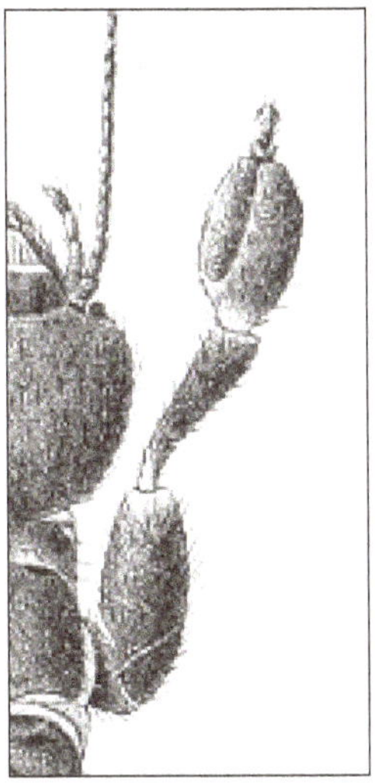

Webspinner, *Embia major*, front leg showing enlarged tarsomere, which contains the silk-spinning organs.

The distal segment of the typical insect leg is the pretarsus. In the Collembola, Protura and many insect larvae, the pretarsus is a single claw. On the pretarsus most insects have a pair of claws unguis, plural ungues. Between the ungues a median unguitractor plate supports the pretarsus. The plate is attached to the apodeme of the flexor muscle of the ungues. In the Neoptera the parempodia are a symmetrical pair of structures arising from the outside (distal) surface of the unguitractor plate between the claws. It is present in many Hemiptera and almost all Heteroptera. Usually the parempodia are bristly (setiform), but in a few species they are fleshy. Sometimes the parempodia are reduced in size so as to almost disappear. Above the unguitractor plate the pretarsus expands forward into a median lobe, the arolium.

Webspinners (Embioptera) have an enlarged basal tarsomere on each of the front legs, containing the silk-producing glands.

Under their pretarsi, members of the Diptera generally have paired lobes or pulvilli, meaning "little cushions". There is a single pulvillus below each unguis. The pulvilli often have an arolium between them or otherwise a median bristle or empodium, meaning the meeting place of the pulvilli. On the underside of the tarsal segments there frequently are pulvillus-like organs or plantulae. The arolium, plantulae and pulvilli are adhesive organs enabling their possessors to climb smooth or steep surfaces. They all are outgrowths of the exoskeleton and their cavities contain blood. Their structures are covered with tubular tenent hairs, the apices of which are moistened by a glandular secretion. The organs are adapted to apply the hairs closely to a smooth surface so that adhesion occurs through surface molecular forces.

Variations in Functional Anatomy of Insect Legs

Bruchid with powerful femora used for escape from hard-shelled seed.

The typical thoracic leg of an adult insect is adapted for running, rather than for digging, leaping, swimming, predation or the like. The legs of most cockroaches are good examples. However, there are many specialized adaptations, including:

- The forelegs of the Gryllotalpidae and some Scarabaeidae are adapted to burrowing in earth.
- The forelegs of the Mantispidae, Mantodea, and Phymatinae are adapted to seizing and holding prey in one way, while those of the Gyrinidae are long and adapted for grasping food or prey in quite a different way.
- The forelegs of some butterflies, such as many Nymphalidae, are reduced so greatly that only two pairs of functional walking legs remain.

- In most Orthoptera the hind legs are saltatorial; they have heavily bipinnately muscled femora and straight, long tibiae adapted to leaping and to some extent to defence by kicking. Flea beetles such as members of the subfamily Halticinae also have powerful hind femora that enable them to leap spectacularly.
- Other beetles with spectacularly muscular hind femora may not be saltatorial at all, but very clumsy; for example, particular species of Bruchinae use their swollen hind legs for forcing their way out of the hard-shelled seeds of plants such as Erythrina in which they grew to adulthood.
- The legs of the Odonata, the dragonflies and damselflies, are adapted for seizing prey that the insects feed on while flying or while sitting still on a plant; they are nearly incapable of using them for walking.
- Hardly any adult aquatic insects use anything but their specially adapted legs for swimming, though many species of immature insects swim by other means, such as by wriggling, undulating, or expelling water.

Evolution and Homology of Arthropod Legs

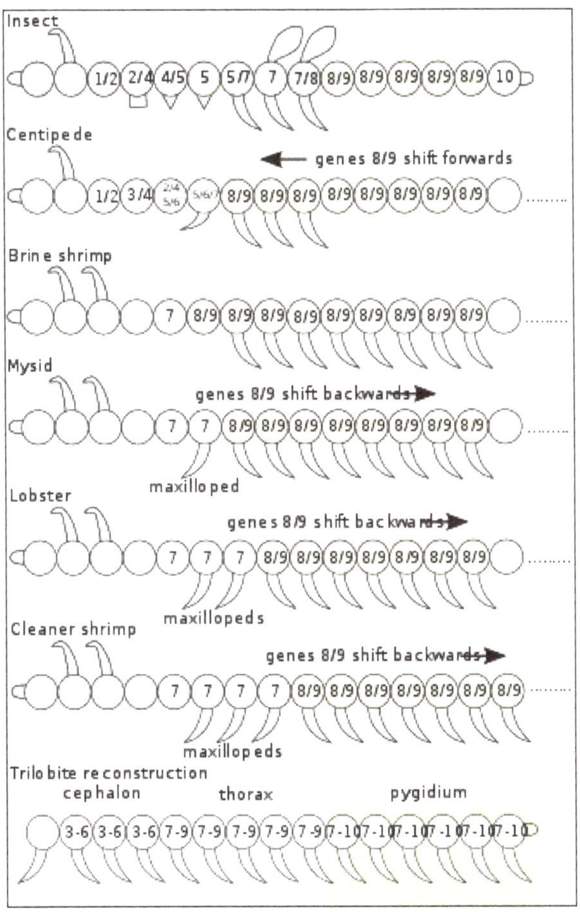

The embryonic body segments (somites) of different arthropods taxa have diverged from a simple body plan with many similar appendages which are serially homologous, into a variety of body plans with fewer segments equipped with specialised appendages. The homologies between these have been discovered by comparing genes in evolutionary developmental biology.

Expression of Hox genes in the body segments of different groups of arthropod, as traced by evolutionary developmental biology. The Hox genes 7, 8, and 9 correspond in these groups but are shifted (by heterochrony) by up to three segments. Segments with maxillopeds have Hox gene 7. Fossil trilobites probably had three body regions, each with a unique combination of Hox genes.

Somite (body segment)	Trilobite (Trilobitomorpha)	Spider (Chelicerata)	Centipede (Myriapoda)	Insect (Hexapoda)	Shrimp (Crustacea)

1	antennae	chelicerae (jaws and fangs)	antennae	antennae	1st antennae
2	1st legs	pedipalps	-	-	2nd antennae
3	2nd legs	1st legs	mandibles	mandibles	mandibles (jaws)
4	3rd legs	2nd legs	1st maxillae	1st maxillae	1st maxillae
5	4th legs	3rd legs	2nd maxillae	2nd maxillae	2nd maxillae
6	5th legs	4th legs	collum (no legs)	1st legs	1st legs
7	6th legs	-	1st legs	2nd legs	2nd legs
8	7th legs	-	2nd legs	3rd legs	3rd legs
9	8th legs	-	3rd legs	-	4th legs
10	9th legs	-	4th legs	-	5th legs

Insects

Except in species in which legs have been lost or become vestigial through evolutionary adaptation, adult insects have six legs, one pair attached to each of the three segments of the thorax. They have paired appendages on some other segments, in particular, mouthparts, antennae and cerci, all of which are derived from paired legs on each segment of some common ancestor.

Some larval insects do however have extra walking legs on their abdominal segments; these extra legs are called prolegs. They are found most frequently on the larvae of moths and sawflies. Prolegs do not have the same structure as modern adult insect legs, and there has been a great deal of debate as to whether they are homologous with them. Current evidence suggests that they are indeed homologous up to a very primitive stage in their embryological development, but that their emergence in modern insects was not homologous between the Lepidoptera and Symphyta. Such concepts are pervasive in current interpretations of phylogeny.

In general the legs of larval insects, particularly in the Endopterygota, vary more than in the adults. As mentioned, some have prolegs as well as "true" thoracic legs. Some have no externally visible legs at all (though they have internal rudiments that emerge as adult legs at the final ecdysis). Examples include the maggots of flies or grubs of weevils. In contrast, the larvae of other Coleoptera, such as the Scarabaeidae and Dytiscidae have thoracic legs, but no prolegs. Some insects that exhibit hypermetamorphosis begin their metamorphosis as planidia, specialised, active, legged larvae, but they end their larval stage as legless maggots, for example the Acroceridae.

Among the Exopterygota the legs of larvae tend to resemble those of the adults in general, except in adaptations to their respective modes of life. For example, the legs of most immature Ephemeroptera are adapted to scuttling beneath underwater stones and the like, whereas the adults have more gracile legs that are less of a burden during flight. Again, the young of the Coccoidea are called "crawlers" and they crawl around looking for a good place to feed, where they settle down and stay for life. Their later instars have no functional legs in most species. Among the Apterygota the legs of immature specimens are in effect smaller versions of the adult legs.

ABDOMEN

An insect's abdomen is the third functional region (tagma) of its body; the abdomen is located just behind the thorax. In most insects, the junction between thorax and abdomen is broad, but in some groups, the junction is very narrow (petiolate) giving the appearance of a "wasp-waist".

Entomologists generally agree that insects arose from primitive arthopod ancestors with eleven-segmented abdomens. Some present-day insects (e.g. silverfish and mayflies) still have all of these segments (or remnants of them), but natural selection in more advanced (or specialized) groups has contributed to a reduction in the number of segments -- sometimes to as few as six or seven (e.g. beetles and flies).

Each segment of the abdomen consists of a dorsal sclerite, the tergum, and a ventral sclerite, the sternum, joined to one another laterally by a pleural membrane. The front margins of each segment often "telecope" inside the sclerites of the preceding sement, allowing the abdomen to expand and contract in response to the actions of skeletal muscles.

In many adult insects, there is a spiracle (opening to the respiratory system) near the pleural membrane on each side of the first eight abdominal segments. Some spiracles may be permanently closed, but still represented by a dimple in the sclerite.

At the very back of the abdomen, the anus (rear opening of the digestive system) is nestled between three protective sclerites: a dorsal epiproct and a pair of lateral paraprocts. A pair of sensory organs, the cerci, may be located near the anterior margin of the paraprocts. These structures are tactile (touch) receptors. They are usually regarded as a "primitive" trait because they are absent in the hemipteroid and holometabolous orders.

The insect's genital opening lies just below the anus: it is surrounded by specialized sclerites that form the external genitalia. In females, paired appendages of the eighth and ninth abdominal segment fit together to form an egg-laying mechanism called the ovipositor. These appendages consist of four valvifers (basal sclerites with muscle attachments) and six valvulae (apical sclerites which guide the egg as it emerges from the female's body). In males, the genital opening is usually enclosed in a tube-like aedeagus which enters the female's body during copulation (like a penis). The external genitalia may also include other sclerites (e.g. subgenital plate, claspers, styli, etc.) that facilitate mating or egg-laying. The structure of these genital sclerites differs from species to species to the extent that it usually prevents inter-species hybridization and also serves as a valuable identification tool for insect taxonomists.

Other abdominal structures may also be present in some insects. These include:

- Pincers: In Dermaptera (earwigs), the cerci are heavily sclerotized and forceps-like. They are used mostly for defense, but also during courtship, and sometimes to help in folding the wings.
- Median caudal filament: A thread-like projection arising from the center of the last abdominal segment (between the cerci). This structure is found only in "primitive" orders (e.g. Diplura, Thysanura, Ephemeroptera).
- Cornicles: Paired secretory structures located dorsally on the abdomen of aphids. The cornicles produce substances that repel predators or elicit care-giving behavior by symbiotic ants.
- Abdominal prolegs: Fleshy, locomotory appendages found only in the larvae of certain orders (notably Lepidoptera, but also Mecoptera and some Hymenoptera).
- Sting: A modified ovipositor, found only in the females of aculeate Hymenoptera (ants, bees, and predatory wasps).
- Abdominal gills: Respiratory organs found in the nymphs (naiads) of certain aquatic insects. In Ephemeroptera (mayflies), paired gills are located along

the sides of each abdominal segment; in Odonata (damselflies), the gills are attached to the end of the abdomen.

- Furcula: The "springtail" jumping organ found in Collembola on the ventral side of the fifth abdominal segment. A clasp (the tenaculum) on the third abdominal segment holds the springtail in its "cocked" position.
- Collophore: A fleshy, peg-like structure found in Collembola on the ventral side of the first abdominal segment. It appears to maintain homeostasis by regulating absorption of water from the environment.

References

- The-insect-thorax: insectsexplained.wordpress.com, Retrieved 23 August, 2020
- Capinera, John L. (2008-08-11). Encyclopedia of Entomology. Springer Science & Business Media. ISBN 9781402062421
- Resh, Vincent H.; Ring T. Carde (July 1, 2009). Encyclopedia of Insects (2 ed.). U. S. A.: Academic Press. ISBN 978-0-12-374144-8
- Gullan, Penny J.; Cranston, Peter S. (2005). The Insects: an Outline of Entomology (3rd ed.). Oxford, UK: Blackwell Publishing. p. 38. ISBN 978-1-4051-1113-3
- Mehlhorn, Heinz (2001). Encyclopedic Reference of Parasitology: Biology, Structure, Function. Springer Science & Business Media. p. 310. ISBN 978-3-540-66819-0
- Triplehorn, Charles A.; Johnson Norman F. (2005). Borror and DeLong's introduction to the study of insects (7th ed.). Thomson Brooks/Cole. ISBN 0-03-096835-6

Internal Anatomy

CHAPTER 4

The internal anatomy of insects consists of the nervous system, digestive system, circulatory system, respiratory system, muscular system, endocrine system and reproductive system. The topics elaborated in this chapter will help in gaining a better perspective of internal anatomy of insects.

Insect has properly defined internal systems which perform specific tasks and supports living of it. Figure shows the "insides" of two omnivorous insects, a cockroach and a cricket, which have relatively unspecialized digestive and reproductive systems. The digestive system, which includes salivary glands as well as an elongate gut, consists of three main sections. These function in storage, biochemical breakdown, absorption, and excretion. Each gut section has more than one physiological role and this may be reflected in local structural modifications, such as thickening of the gut wall or diverticula (extensions) from the main lumen. The reproductive systems depicted in Figure. exemplify the female and male organs of many insects. These may be dominated in males by very visible accessory glands, especially as the testes of many adult insects are degenerate or absent. This is because the spermatozoa are produced in the pupal or penultimate stage and stored. In gravid female insects, the body cavity may be filled with eggs at various stages of development, thereby obscuring other internal organs. Likewise, the internal structures (except the gut) of a well-fed, late-stage caterpillar may be hidden within the mass of fat body tissue.

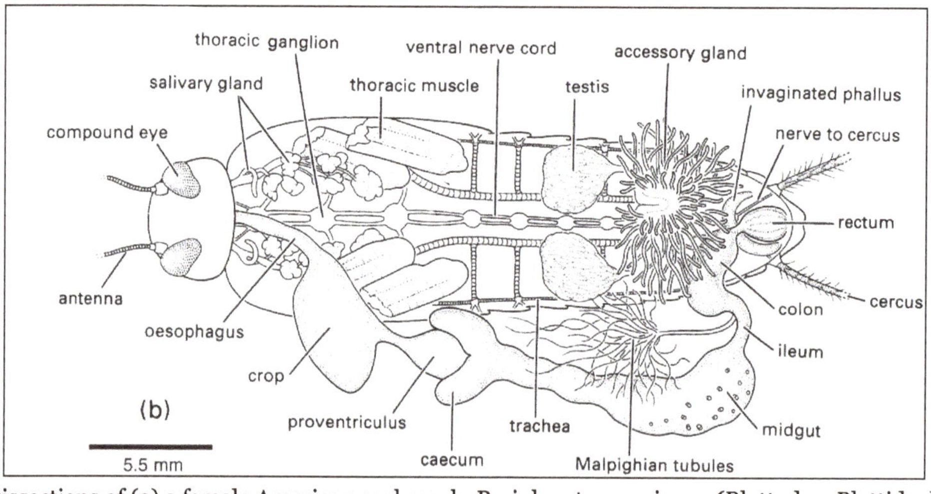

Dissections of (a) a female American cockroach, Periplaneta americana (Blattodea: Blattidae).

The insect body cavity, called the hemocoel (haemocoel) and filled with fluid hemolymph (haemolymph), is lined with endoderm and ectoderm. It is not a true coelom, which is defined as a mesoderm-lined cavity. Hemolymph (so-called because it combines many roles of vertebrate blood (hem/haem) and lymph) bathes all internal organs, delivers nutrients, removes metabolites, and performs immune functions. Unlike vertebrate blood, hemolymph rarely has respiratory pigments and therefore has little or no role in gaseous exchange. In insects this function is performed by the tracheal system, a ramification of air-filled tubes (tracheae), which sends fine branches throughout the body. Gas entry to and exit from tracheae is controlled by sphincter-like structures called spiracles that open through the body wall. Non-gaseous wastes are filtered from the hemolymph by filamentous Malpighian tubules (named after their discoverer), which have free ends distributed through the hemocoel. Their contents are emptied into the gut from which, after further modification, wastes are eliminated eventually via the anus.

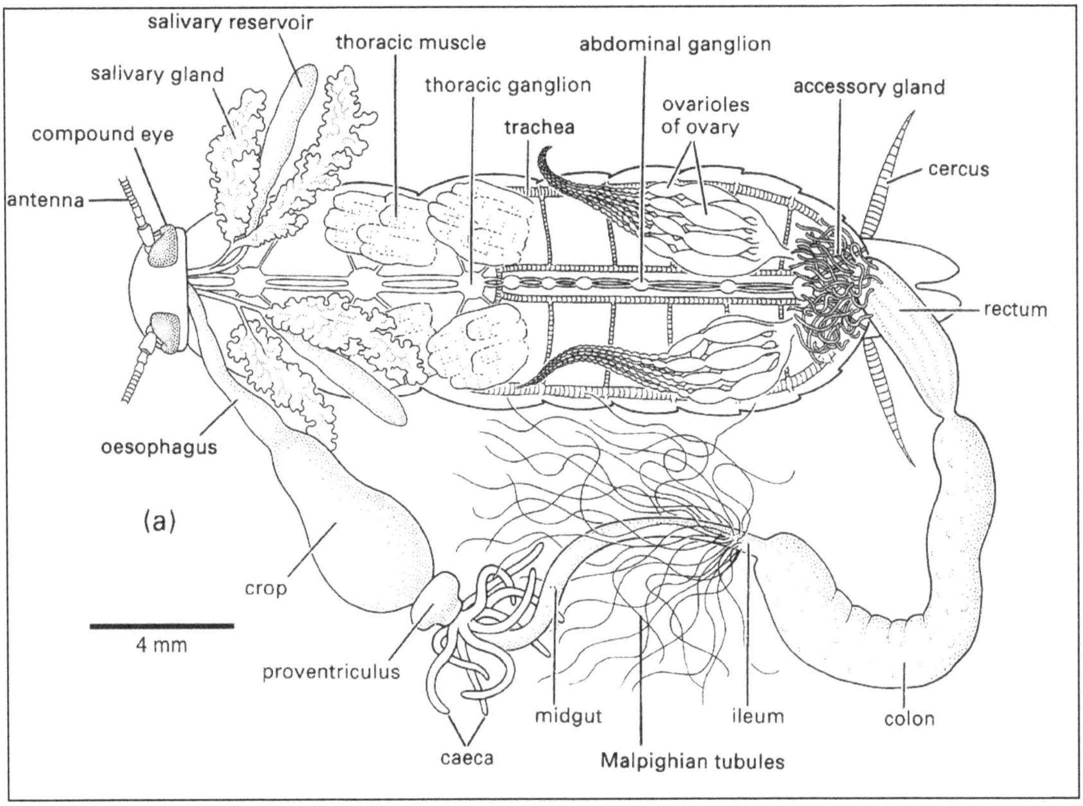

(b) a male black field cricket, Teleogryllus commodus (Orthoptera: Gryllidae).

All motor, sensory, and physiological processes in insects are controlled by the nervous system in con- junction with hormones (chemical messengers). The brain and ventral nerve cord are readily visible in dissected insects, but most endocrine centers, neurosecretion sites, numerous nerve fibers, muscles, and other tissues cannot be seen by the unaided eye.

NERVOUS SYSTEM

Insects have a complex nervous system which incorporates a variety of internal physiological information as well as external sensory information. As in the case of vertebrates, the basic component is the neuron or nerve cell. This is made up of a dendrite with two projections that receive stimuli and an axon, which transmits information to another neuron or organ, like a muscle. As for vertebrates, chemicals (neurotransmitters such as acetylcholine and dopamine) are released at synapses.

Central Nervous System

An insect's sensory, motor and physiological processes are controlled by the central nervous system along with the endocrine system. Being the principal division of the nervous system, it consists of a brain, a ventral nerve cord and a subesophageal ganglion which is connected to the brain by two nerves, extending around each side of the oesophagus.

The brain has three lobes:

- Protocerebrum, innervating the compound eyes and the ocelli.
- Deutocerebrum, innervating the antennae.
- Tritocerebrum, innervating the foregut and the labrum.

The ventral nerve cord extends from the suboesophageal ganglion posteriorly. A layer of connective tissue called the neurolemma covers the brain, ganglia, major peripheral nerves and ventral nerve cords.

The head capsule (made up of six fused segments) has six pairs of ganglia. The first three pairs are fused into the brain, while the three following pairs are fused into the subesophageal ganglion. The thoracic segments have one ganglion on each side, which are connected into a pair, one pair per segment. This arrangement is also seen in the abdomen but only in the first eight segments. Many species of insects have reduced numbers of ganglia due to fusion or reduction. Some cockroaches have just six ganglia in the abdomen, whereas the wasp *Vespa crabro* has only two in the thorax and three in the abdomen. And some, like the house fly *Musca domestica*, have all the body ganglia fused into a single large thoracic ganglion. The ganglia of the central nervous system act as the coordinating centres with their own specific autonomy where each may coordinate impulses in specified regions of the insect's body.

Peripheral Nervous System

This consists of motor neuron axons that branch out to the muscles from the ganglia of the central nervous system, parts of the sympathetic nervous system and the sensory neurons of the cuticular sense organs that receive chemical, thermal, mechanical or visual stimuli from the insects environment. The sympathetic nervous system includes nerves and the ganglia that innervate the gut both posteriorly and anteriorly, some endocrine organs, the spiracles of the tracheal system and the reproductive organs.

Sensory Organs

Chemical senses include the use of chemoreceptors, related to taste and smell, affecting mating, habitat selection, feeding and parasite-host relationships. Taste is usually located on the mouthparts of the insect but in some insects, such as bees, wasps and ants, taste organs can also be found on the antennae. Taste organs can also be found on the tarsi of moths, butterflies and flies. Olfactory sensilla enable insects to smell and are usually found in the antennae. Chemoreceptor sensitivity related to smell in some substances, is very high and some insects can detect particular odours that are at low concentrations miles from their original source.

Mechanical senses provide the insect with information that may direct orientation, general movement, flight from enemies, reproduction and feeding and are elicited from the sense organs that are sensitive to mechanical stimuli such as pressure, touch and vibration. Hairs (setae) on the cuticle are responsible for this as they are sensitive to vibration touch and sound.

Hearing structures or tympanal organs are located on different body parts such as, wings, abdomen, legs and antennae. These can respond to various frequencies ranging from 100 Hz to 240 kHz depending on insect species. Many of the joints of the insect have tactile setae that register movement. Hair beds and groups of small hair like sensilla, determine proprioreception or information about the position of a limb, and are found on the cuticle at the joints of segments and legs. Pressure on the body wall or strain gauges are detected by the campiniform sensilla and internal stretch receptors sense muscle distension and digestive system stretching.

The compound eye and the ocelli supply insect vision. The compound eye consists of individual light receptive units called ommatidia. Some ants may have only one or two, however dragonflies may have over 10,000. The more ommatidia the greater the visual acuity. These units have a clear lens system and light sensitive retina cells. By day, the image flying insects receive is made up of a mosaic of specks of differing light intensity from all the different ommatidia. At night or dusk, visual acuity is sacrificed for light sensitivity. The ocelli are unable to form focused images but are sensitive mainly, to differences in light intensity. Colour vision occurs in all orders of insects. Generally insects see better at the blue end of the spectrum than at the red end. In some orders sensitivity ranges can include ultraviolet.

A number of insects have temperature and humidity sensors and insects being small, cool more quickly than larger animals. Insects are generally considered cold-blooded or ectothermic, their body temperature rising and falling with the environment. However, flying insects raise their body temperature through the action of flight, above environmental temperatures.

The body temperature of butterflies and grasshoppers in flight may be 5 °C or 10 °C above environmental temperature, however moths and bumblebees, insulated by scales and hair, during flight, may raise flight muscle temperature 20–30 °C above the environment temperature. Most flying insects have to maintain their flight muscles above a certain temperature to gain power enough to fly. Shivering, or vibrating the wing muscles allow larger insects to actively increase the temperature of their flight muscles, enabling flight.

Until very recently, no one had ever documented the presence of nociceptors (the cells that detect and transmit sensations of pain) in insects, though recent findings of nociception in larval fruit flies challenges this and raises the possibility that some insects may be capable of feeling pain.

Nociceptor

A nociceptor ("pain receptor") is a sensory neuron that responds to damaging or potentially damaging stimuli by sending "possible threat" signals to the spinal cord and the brain. If the brain perceives the threat as credible, it creates the sensation of pain to direct attention to the body part, so the threat can hopefully be mitigated; this process is called nociception.

Types and Functions

The peripheral terminal of the mature nociceptor is where the noxious stimuli are detected and transduced into electrical energy. When the electrical energy reaches a threshold value, an action potential is induced and driven towards the central nervous system (CNS). This leads to the train of events that allows for the conscious awareness of pain. The sensory specificity of nociceptors is established by the high threshold only to particular features of stimuli. Only when the high threshold has been reached by either chemical, thermal, or mechanical environments are the nociceptors triggered. The majority of nociceptors are classified by which of the environmental modalities they respond to. Some nociceptors respond to more than one of these modalities and are consequently designated polymodal. Other nociceptors respond to none of these modalities (although they may respond to stimulation under conditions of inflammation) and are referred to as sleeping or silent.

Nociceptors have two different types of axons. The first are the Aδ fiber axons. They are myelinated and can allow an action potential to travel at a rate of about 20 meters/second towards the CNS. The other type is the more slowly conducting C fiber axons. These only conduct at speeds of around 2 meters/second. This is due to the light or non-myelination of the axon. As a result, pain comes in two phases. The first phase is mediated by the fast-conducting Aδ fibers and the second part due to (Polymodal) C fibers. The pain associated with the Aδ fibers can be associated to an initial extremely sharp pain. The second phase is a more prolonged and slightly less intense feeling of pain as a result of the acute damage. If there is massive or prolonged input to a C fiber, there is a progressive build up in the spinal cord dorsal horn; this phenomenon is similar to tetanus in muscles but is called wind-up. If wind-up occurs there is a probability of increased sensitivity to pain.

Thermal

Thermal nociceptors are activated by noxious heat or cold at various temperatures. There are specific nociceptor transducers that are responsible for how and if the specific nerve ending responds to the thermal stimulus. The first to be discovered was TRPV1, and it has a threshold that coincides with the heat pain temperature of 42 °C. Other temperature in the warm–hot range is mediated by more than one TRP channel. Each of these channels express a particular C-terminal domain that corresponds to the warm–hot sensitivity. The interactions between all these channels and how the temperature level is determined to be above the pain threshold are unknown at this time. The cool stimuli are sensed by TRPM8 channels. Its C-terminal domain differs from the heat sensitive TRPs. Although this channel corresponds to cool stimuli, it is still unknown whether it also contributes in the detection of intense cold. An interesting finding related to cold stimuli is that tactile sensibility and motor function deteriorate while pain perception persists.

Mechanical

Mechanical nociceptors respond to excess pressure or mechanical deformation. They also respond to incisions that break the skin surface. The reaction to the stimulus is processed as pain by the cortex, just like chemical and thermal responses. These mechanical nociceptors frequently have polymodal characteristics. So it is possible that some of the transducers for thermal stimuli are the same for mechanical stimuli. The same is true for chemical stimuli, since TRPA1 appears to detect both mechanical and chemical changes.

Chemical

Chemical nociceptors have TRP channels that respond to a wide variety of spices. The one that sees the most response and is very widely tested is capsaicin. Other chemical stimulants are environmental irritants like acrolein, a World War I chemical weapon and a component of cigarette smoke. Apart from these external stimulants, chemical nociceptors have the capacity to detect endogenous ligands, and certain fatty acid amines that arise from changes in internal tissues. Like in thermal nociceptors, TRPV1 can detect chemicals like capsaicin and spider toxins.

Sleeping/Silent

Although each nociceptor can have a variety of possible threshold levels, some do not respond at all to chemical, thermal or mechanical stimuli unless injury actually has occurred. These are typically referred to as silent or sleeping nociceptors since their response comes only on the onset of inflammation to the surrounding tissue.

Polymodal

Many neurons perform only a single function; therefore, neurons that perform these functions in combination are given the classification "polymodal."

DIGESTIVE SYSTEM

An insect uses its digestive system to extract nutrients and other substances from the food it consumes. Most of this food is ingested in the form of macromolecules and other complex substances (such as proteins, polysaccharides, fats, nucleic acids, etc.) which must be broken down by catabolic reactions into smaller molecules (i.e. amino acids, simple sugars, etc.) before being used by cells of the body for energy, growth, or reproduction. This break-down process is known as digestion.

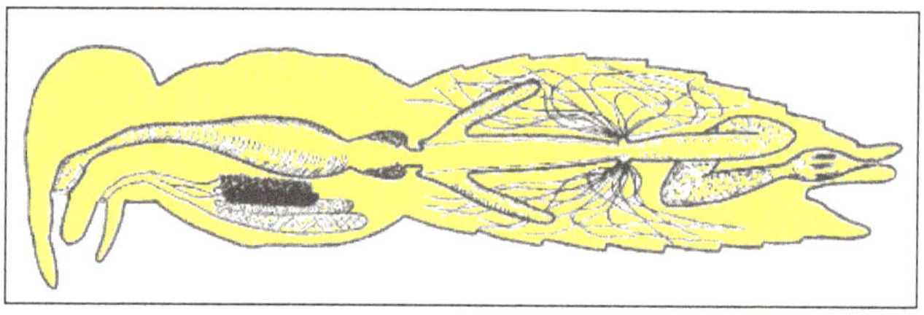

All insects have a complete digestive system. This means that food processing occurs within a tube-like enclosure, the alimentary canal, running lengthwise through the body from mouth to anus. Ingested food usually travels in only one direction. This arrangement differs from an incomplete digestive system (found in certain lower invertebrates like hydra and flatworms) where a single opening to a pouch-like cavity serves as both mouth and anus. Most biologists regard a complete digestive system as an evolutionary improvement over an incomplete digestive system because it permits functional specialization – different parts of the system may be specially adapted for various functions of food digestion, nutrient absorption, and waste excretion. In most insects, the alimentary canal is subdivided into three functional regions: foregut (stomodeum), midgut (mesenteron), and hindgut (proctodeum).

In addition to the alimentary canal, insects also have paired salivary glands and salivary reservoirs. These structures usually reside in the thorax (adjacent to the foregut). Salivary ducts lead from the glands to the reservoirs and then forward, through the head, to an opening (the salivarium) behind the hypopharynx. Movements of the mouthparts help mix saliva with food in the buccal cavity.

Stomodaeum

An insect's mouth, located centrally at the base of the mouthparts, is a muscular valve (sphincter) that marks the "front" of the foregut. Food in the buccal cavity is sucked through the mouth opening and into the pharynx by the action of cibarial muscles. These muscles are located between the head capsule and the anterior wall of the pharynx. When they contract, they create suction by enlarging the volume of the pharynx (like opening a bellows). This "suction pump" mechanism is called the cibarial pump. It is especially well-developed in insects with piercing/sucking mouthparts. From the pharynx, food passes into the esophagus by means of peristalsis (rhythmic muscular contractions of the gut wall). The esophagus is just a simple tube that connects the pharynx to the crop, a food-storage organ.

Food remains in the crop until it can be processed through the remaining sections of the alimentary canal. While in the crop, some digestion may occur as a result of salivary enzymes that were added in the buccal cavity and/or other enzymes regurgitated from the midgut. In some insects, the crop opens posteriorly into a muscular proventriculus. This organ contains tooth-like denticles that grind and pulverize food particles. The proventriculus serves much the same function as a gizzard in birds. The stomodeal valve, a sphincter muscle located just behind the proventriculus, regulates the flow of food from the stomodeum to the mesenteron.

In a developing embryo, the foregut arises as a simple invagination of the anterior body wall: this means that all of its tissues and organs are derived from embryonic ectoderm. In effect, the inside of the stomodeum is continuous with the outside of the insect's body. Since exoskeleton is secreted to protect the insect externally, it is not surprising to find that cells lining the foregut produce a similar structure (known as the intima) to protect themselves from abrasion by food particles. The hard denticles inside the proventriculus are made from this same material.

Mesenteron

The midgut begins just past the stomodeal valve. Near its anterior end, finger-like projections (usually from 2 to 10) diverge from the walls of the midgut. These structures, the gastric caecae, provide extra surface area for secretion of enzymes or absorption of water (and other substances) from the alimentary canal. The rest of the midgut is called the ventriculus – it is the primary site for enzymatic digestion of food and absorption of nutrients. Digestive cells lining the walls of the ventriculus have microscopic projections (microvilli) that increase surface area for nutrient absorption.

The midgut is derived from embryonic endoderm so it is not protected by an intima. Instead, the midgut is lined with a semipermeable membrane secreted by a cluster of cells (the cardial epithelium) that lie just behind the stomodeal valve. This peritrophic membrane consists of chitin fibrils embedded in a protein-carbohydrate matrix. It protects the delicate digestive cells without inhibiting absorption of nutrient molecules.

The posterior end of the midgut is marked by another sphincter muscle, the pyloric valve. It regulates the flow of material from the mesenteron into the proctodaeum.

Proctodaeum

The pyloric valve serves as a point of origin for dozens to hundreds of Malpighian tubules. These long, spaghetti-like structures extend throughout most of the abdominal cavity where they serve as excretory organs, removing nitrogenous wastes (principally ammonium ions, NH_4^+) from the hemolymph. The toxic NH_4^+ is quickly converted to urea and then to uric acid by a series of chemical reactions within the Malpighian tubules. The uric acid, a semi-solid, accumulates inside each tubule and is eventually emptied into the hindgut for elimination as part of the fecal pellet.

The rest of the hindgut plays a major role in homeostasis by regulating the absorption of water and salts from waste products in the alimentary canal. In some insects, the hindgut is visibly subdivided into an ileum, a colon, and a rectum. Efficient recovery of water is facilitated by six rectal pads that are embedded in the walls of the rectum. These organs remove more than 90% of the water from a fecal pellet before it passes out of the body through the anus.

Embryonically, the hindgut develops as an invagination of the body wall (from ectodermal tissue). Just like the foregut, it is lined with a thin, protective layer of cuticle (intima) that is secreted by the endothelial cells of the gut wall. When an insect molts, it sheds and replaces the intima in both the foregut and the hindgut.

CIRCULATORY SYSTEM

Insects, like all other arthropods, have an open circulatory system which differs in both structure and function from the closed circulatory system found in humans and other vertebrates. In a closed system, blood is always contained within vessels (arteries, veins, capillaries, or the heart itself). In an open system, blood (usually called hemolymph) spends much of its time flowing freely within body cavities where it makes direct contact with all internal tissues and organs.

The circulatory system is responsible for movement of nutrients, salts, hormones, and metabolic wastes throughout the insect's body. In addition, it plays several critical roles in defense: it seals off wounds through a clotting reaction, it encapsulates and destroys internal parasites or other invaders, and in some species, it produces (or sequesters) distasteful compounds that provide a degree of protection against predators. The hydraulic (liquid) properties of blood are important as well. Hydrostatic pressure generated internally by muscle contraction is used to facilitate hatching, molting, expansion of body and wings after molting, physical movements (especially in soft-bodied larvae), reproduction (e.g. insemination and oviposition), and evagination of certain types of exocrine glands. In some insects, the blood aids in thermoregulation: It can help cool the body by conducting excess heat away from active flight muscles or it can warm the body by collecting and circulating heat absorbed while basking in the sun.

A dorsal vessel is the major structural component of an insect's circulatory system. This tube runs longitudinally through the thorax and abdomen, along the inside of the dorsal body wall. In most insects, it is a fragile, membranous structure that collects hemolymph in the abdomen and conducts it forward to the head.

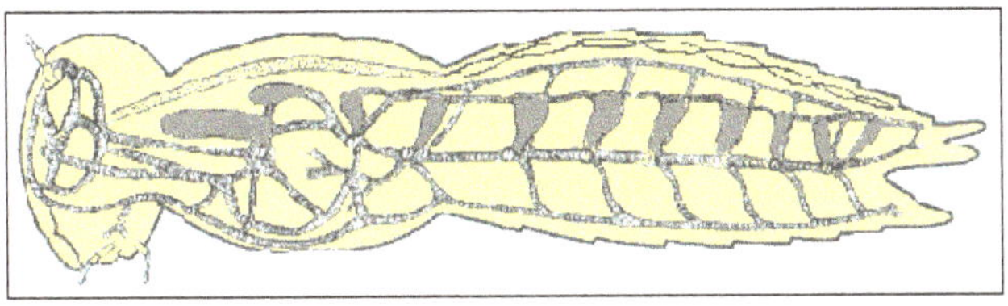

In the abdomen, the dorsal vessel is called the heart. It is divided segmentally into chambers that are separated by valves (ostia) to ensure one-way flow of hemolymph. A pair of alary musclesare attached laterally to the walls of each chamber. Peristaltic contractions of the these muscles force the hemolymph forward from chamber to chamber. During each diastolic phase (relaxation), the ostia open to allow inflow of hemolymph from the body cavity. The heart's contraction rate varies considerably from species to species — typically in the range of 30 to 200 beats per minute. The rate tends to fall as ambient temperature drops and rise as temperature (or the insect's level of activity) increases.

In front of the heart, the dorsal vessel lacks valves or musculature. It is a simple tube (called the aorta) which continues forward to the head and empties near the brain. Hemolymph bathes the organs and muscles of the head as it emerges from the aorta, and then haphazardly percolates back over the alimentary canal and through the body until it reaches the abdomen and re-enters the heart.

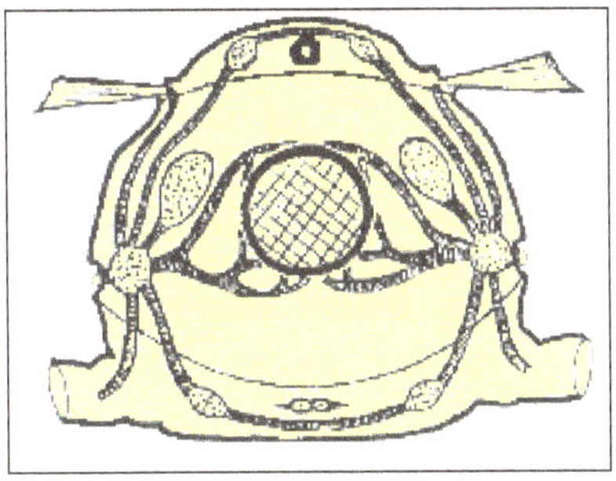

To facilitate circulation of hemolymph, the body cavity is divided into three compartments (called blood sinuses) by two thin sheets of muscle and/or membrane known as the dorsal and ventral diaphragms. The dorsal diaphragm is formed by alary muscles of the heart and related structures; it separates the pericardial sinus from the perivisceral sinus. The ventral diaphragm usually covers the nerve cord; it separates the perivisceral sinus from the perineural sinus.

In some insects, pulsatile organs are located near the base of the wings or legs. These muscular "pumps" do not usually contract on a regular basis, but they act in conjunction with certain body movements to force hemolymph out into the extremities.

About 90% of insect hemolymph is plasma: a watery fluid — usually clear, but sometimes greenish or yellowish in color. Compared to vertebrate blood, it contains relatively high concentrations of amino acids, proteins, sugars, and inorganic ions. Overwintering insects often sequester enough ribulose, trehalose, or glycerol in the plasma to prevent it from freezing during the coldest winters. The remaining 10% of hemolymph volume is made up of various cell types (collectively known as hemocytes); they are involved in the clotting reaction, phagocytosis, and/or encapsulation of foreign bodies. The density of insect hemocytes can fluctuate from less than 25,000 to more than 100,000 per cubic millimeter, but this is significantly fewer than the 5 million red blood cells, 300,000 platelets, and 7000 white blood cells found in the same volume of human blood. With the exception of a few aquatic midges, insect hemolymph does NOT contain hemoglobin (or red blood cells). Oxygen is delivered by the tracheal system, not the circulatory system.

RESPIRATORY SYSTEM

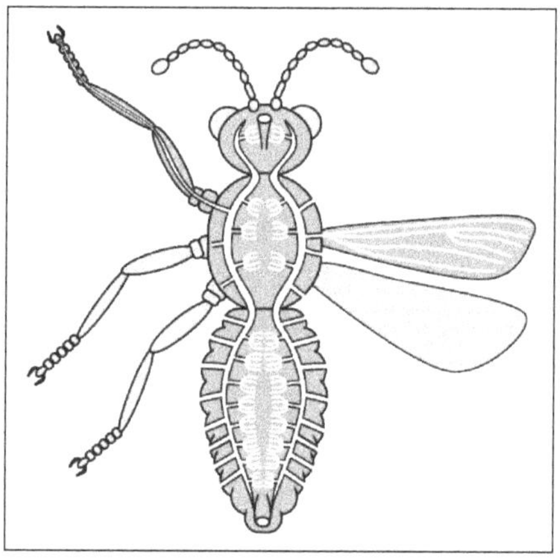

The basic insect respiratory system consists of a series of rigid tubes, called tracheae (singular trachea), connected to the outside via pairs of pores called spiracles (typically one pair per segment on the sides of the thorax and abdomen, lacking on certain segments). Air enters the system via the spiracles and the tracheae are air-filled. The spiracles can often be opened and closed and lead into short tracheae that enter a pair of longitudinal tracheal trunks, which are the main tracheal tubes. From these lateral tracheae branch smaller tracheae that supply the tissues with air. This supply is especially rich in the more active tissues, such as muscles, nervous tissues and the gut. Tracheae also extend into the wings, running inside the wing veins. The tracheae branch until they reach adiameter of 2 to 5 micrometres (2-5 thousands of a millimetre) and then often enter stellate tracheole cells (transition cells) from which they emerge as finer branches called tracheoles, with diameters less than one micrometre. These tracheoles terminate inside the tissues, almost always as open-ended or blind-ended tubes about 200 nanometres in diameter (200 millionths of a millimetre).

The outside cuticle of the insect extends inwards through the spiracles to line the inside of the main tracheae and in the smaller tracheae this cuticle is reduced to a thin membrane lining the lumen of the tracheae and forming the taenidia. The tracheoles are often said to lack this tough inner lining, but end. Apart from this lining, the walls of the tracheae consist of a single layer of flattened epithelial cells.

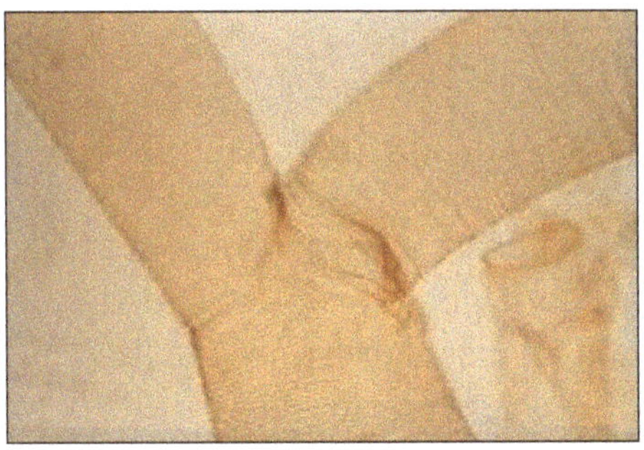

A branch in a trachea from *Bombyx mori*.

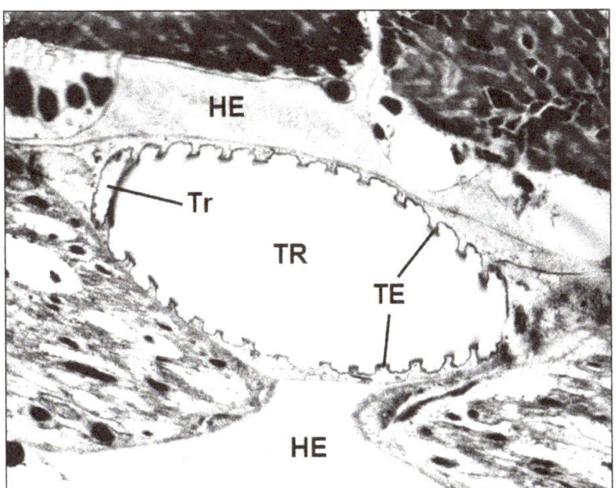

A cross-section through a trachea (TR) in the antenna of the rove beetle *Aleochara bilineata*, illustrating the taenidia (TE) and a small tracheole (Tr) branching off. The trachea is supplying nerves and muscles and is bathed in haemolymph (HE).

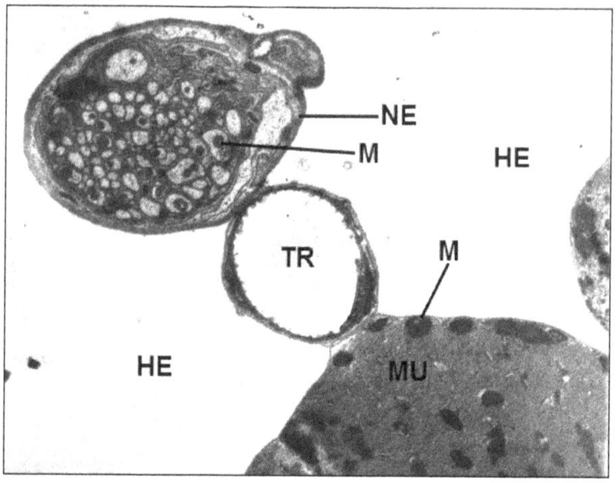

A tracheole (TR) in the antenna of Aleochara bilineata, supplying a muscle (MU) and a nerve (NE). M: Mitochondrion.

The tracheoles may terminate on the surface of a cell, such as a muscle cell, or they may penetrate inside the cell, either part way or even forming an extensive network inside and also covering the outside of the muscle. The supply is generally greater to flight muscles, especially of the fibrillar type.

The Role of Fluid in the Tracheoles

When an insect is at rest the ends of the tracheoles are filled with fluid. However, oxygen diffuses faster in air than it does in water, and the fluid is actually a barrier to oxygen diffusion. Thus, when an insect exercises the fluid gets 'sucked' into the muscle cell until oxygen reaches the ends of the tracheoles. (This happens, at least in part, as the concentration of solutes build up in the exercising muscle, drawing in the water by osmosis). The fluid is there at rest because the tissues are bathed in fluid, though it may serve to reduce water-loss and dehydration through the tracheal system. When an insect hatches from its egg, its tracheal system is initially filled with fluid, but this fluid is actively absorbed by the tracheole cells until the system fills with air.

The Transport of Air through the Tracheal System

In the basic system described so far, air simply diffuses in through the spiracles and along the tracheae. Diffusion is rapid over a millimetre or even a centimetre, but is very slow at greater distances. This diffusion-driven system occurs in some small or not very active insects. In large and active insects, such as moths, butterflies, bees and wasps, diffusion alone is insufficient. These latter insects show breathing movements - that is they actively pump air through the tracheal system. This is why the abdomen pulses in these insects. Sometimes only the tergum of each abdominal segment moves up and down, as in beetles, or both the sternum and tergum, as in flies, or the side-walls (pleura) may be very flexible and also move in and out, greatly changing the internal volume of the abdomen, a sin moths and butterflies. In this way a rapid stream of air flows through the tracheal system. Certain spiracles may be used to take air in, others to expel air, e.g. air may be drawn in through the thorax and expelled through the abdomen. However, these circuits are not hard and fast and occasionally the direction of flow may be reversed.

Air sacs may facilitate this movement of air through the tracheae. Air sacs can occur in almost any part of the system, and in rigid structures like the head and thorax they may be permanently expanded, acting as reservoirs of air, whilst in the abdomen they may greatly inflate and contract (flatten and empty). The diagrams below illustrate the air sacs in the abdomen of a honey-bee worker. You may have noticed how the abdomen of a honey-bee pulsates as the muscles of the abdomen expand and contract the abdominal segments to fill and deflate the air sacs. If you have ever chased a bee or wasp, then you may also have noticed that the abdomen pulsates harder and faster with exertion. These breathing movements are intimately connected to the circulatory system of the insect also.

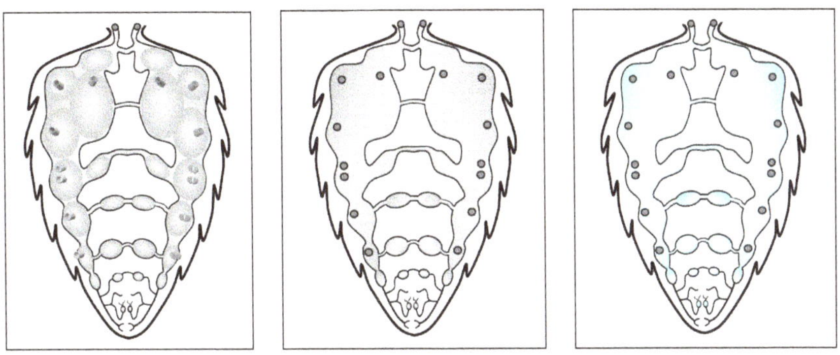

Above: Air sacs in the abdomen of a honey-bee worker. The tubes (or circles) indicate the main tracheae entering/leaving the air sacs.

Control of Spiracle Opening and Closing

Insects lose most of their water through the spiracles. Being small they do not have much water to lose! It is not surprising, therefore, that insects typically only open their spiracles when they need more oxygen. An increase in carbon dioxide, a product of respiration, causes the spiracles to open more. Starvation and reduction in metabolic rate causes them to opening at the end of the trachea, opening at the end of the trachea, rather than the outermost opening of the rather than the outermost opening of the that is the internal opening of the atrium, that is the internal opening of the atrium, atrium (a naming convention I personally find unhelpful since in insects lacking atria, the spiracle opens directly to the outside and in diagrams the external openings are generally labeled as 'spiracles' regardless and so I find myself referring to the most external opening, or even the whole structure, as a spiracle, atrium or not).The external opening of the atrium is often screened by spines, meshes and similar structures may cover the atrial openings in some insects, especially those from dry conditions, though both the spines and the atrium may trap water moisture, they may actually be more important in preventing dust from entering the tracheae. An internal spiracular valve often occurs between the spiracle and atrium, which can open or close the spiracle.

Temperature is an important factor. At low temperatures, when metabolism may be low, the spiracles are largely closed but open occasionally. At higher temperatures they may open and close periodically, and at still higher temperatures they may open continuously.

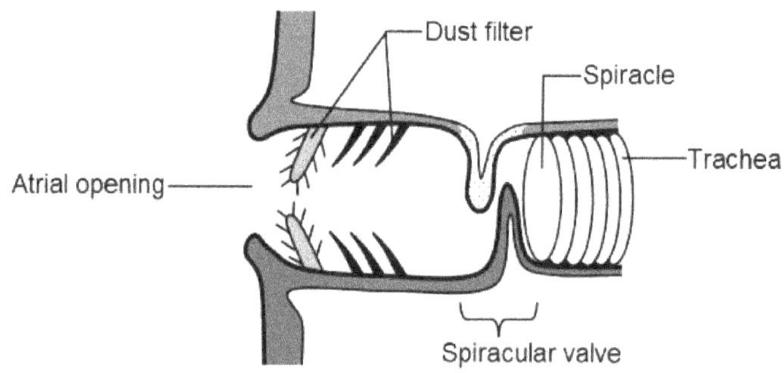

A spiracle equipped with valve, atrium and dust filter

Respiration in Aquatic Insects

Insects have to be able to obtain oxygen if they are to survive when submerged under water. Some insects have what is called a plastron mechanism - the hairs on their bodies are specially modified to trap a film or bubble of air when they dive (these hairs are designed to resist collapse under high pressures). In Dytiscus (Great Diving Beetle) and Notonecta (Water Boatman) air is trapped beneath the elytra (wing covers) and this air communicates directly with the insect's spiracles (openings to its airways or tracheae). These plastrons and sub-elytral air-spaces can function as what are called 'physical gills' - trapped pockets of air that can actually absorb more oxygen from the surrounding water. Oxygen diffuses across an air/water interface some three times faster than nitrogen, so as the insect takes up the oxygen from the trapped air bubble, lowering the partial pressure of oxygen and raising that of nitrogen within the bubble, oxygen diffuses in (when the partial pressure of oxygen in the water exceeds that in the air space) faster than nitrogen diffuses out. This maintains the air bubble for longer and in experiments in which the air was replaced by pure oxygen, the insect actually had to resurface for fresh oxygen sooner than when air was used, since with oxygen the partial pressure of oxygen in the bubble is always greater than in the surrounding water and although the insect takes more oxygen with it to begin with, it is unable to extract any from the surrounding water. In Notonecta (the Water Boatman) the hindlegs are used to drive water currents over the physical gills to irrigate them with fresh oxygenated water. Eventually all the nitrogen in the air space dissolves and then the insect must resurface to replenish its supply of air. In Dytiscus and Notonecta the trapped air volume can be regulated when under water, acting to regulate buoyancy.

Structures that have a higher affinity for air than for water, like plastron hairs, are called hydrofuge structures. Hydrofuge hairs also exist on the siphons of mosquito larvae. Some aquatic insect larvae, like mosquitoes, breath through a siphon - a tube with a spiracle at its tip that leads straight into the tracheal airways inside the insect. The spiracles are surrounded by hydrofuge hairs which repel water and prevent it from entering the system and drowning the insect. The water trough in the neighbouring meadow has dozens of these larvae that wriggle for cover whenever one's shadow passes over them. The hydrofuge hairs cannot repel oil, which is used to control mosquitoes by applying a film of oil to the water's surface; when the larvae try to breathe the oil enters through the spiracle and they drown. Eristalis, a type of hoverfly, has aquatic larvae that are called rat-tailed maggots because of the long tail-like extensible breathing siphon.

Some aquatic insects have a closed tracheal system which does not open to the outside air via spiracles. These include the dragonfly nymphs which have 6 double rows of lamellae lining the rectum and forming the branchial basket. Water is pumped over these tracheal gills, mostly in and out through the anus, and oxygen diffuses across to the trachea which fill the gills. Some aquatic insects have spiracular gills, like *Simulium* (black fly) pupae.

Bloodworms are red worm-like aquatic insect larvae that are often found at the bottom of ponds. These are the larvae of midges, like Chironomus, and certain flies (belonging to the true-flies or Diptera). They are red because they contain a form of haemoglobin (which contains 2 haem groups per molecule instead of 4 as are found in vertebrate haemoglobin) in their haemolymph ('blood plasma'). This haemoglobin is used to store oxygen, enough for two days in low oxygen conditions (as might occur in warm stagnant water). When oxygen is adequate they return to tracheal respiration and replenish their oxygen reserves. The larva of Chironomus has a closed tracheal system and absorbs oxygen across its 'skin' (cuticle) and also has so-called 'blood gills'. These are regions of the body wall that are very thin and project from the body surface as blood-filled sacs that are more-or-less devoid of tracheae. However, these do not seem to have a normal respiratory function, but may assist in recovery from oxygen starvation. The larvae of some mosquitoes have long anal papillae (projections) filled with tracheae and which are held in a current of water created by mouth brushes. However, the respiratory role of such tracheal-filled appendages is hard to determine experimentally and a normal respiratory function is doubted, though they may serve to excrete carbon dioxide.

Aquatic larvae may have closed tracheal systems, with no spiracles, or open tracheal systems, connected to the outside via spiracles. In the former, oxygen is absorbed through the cuticle, into the tracheal system (cuticular respiration). This may be facilitated by tracheal gills - structures filled with highly-branched, often feathery, tracheae that absorb oxygen across the cuticle. Often these gills are folds in the wall of the rectum (rectal gills) in which case the rectum may actively pump water across them, or they may be external appendages filled with tracheae. Sometimes these appendages are very long and bizarre looking, but have sometimes been shown to have no gill function, as their removal may have no effect on oxygen uptake, in which case they may function only in emergencies, or as an oxygen store, or they possibly act as flotation devices. In those larvae with functional spiracles, often at one end of the animal, the fluid is only actively removed from the system once the insect reaches the surface and starts taking in air, often through a snorkel like siphon bearing the spiracles, as in mosquitoes.

Cuticular respiration is often still significant in terrestrial insects. Some butterflies absorb a significant proportion of their oxygen through the large surface area of their membranous wings. Generally, however, little oxygen is absorbed across the skin of insects.

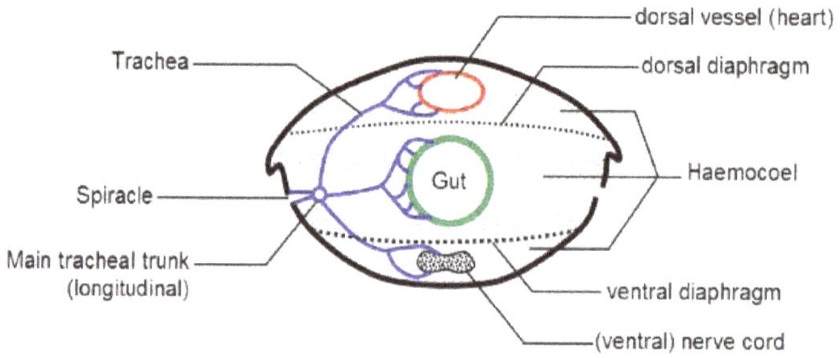

Transverse section of insect abdomen
(Showing tracheal system on left-side only)

MUSCULAR SYSTEM

Muscle is the excitable, contractile tissue of animals that is responsible for movement and behavior. Although there is great variability in structure and performance among different insect muscles, many basic features of biochemical composition, ultrastructural organization, and contractile performance are common among insect muscles and indeed are similar between muscles of insects and those of vertebrates.

Muscle Structure and Ultrastructure

Muscle Fibers and Fiber Bundles

The skeletal muscles of insects are bundles of elongate, multinu-cleate cells called muscle fibers. The fibers attach at each end to the exoskeleton. The muscles typically span joints of the exoskeleton and, when active, cause bending of the joint or stabilization of the joint against external forces. Skeletal muscles are the muscles of behavior, the muscles involved in posture and locomotion. In addition to skeletal muscles, insects contain visceral muscles that cause movement of the gut, Malpighian tubules, and parts of the reproductive system; there are also cardiac muscles that cause contraction of tissue sheets and vessels associated with the circulatory system. The visceral and cardiac muscle cells are typically small, spindle shaped, and with a single nucleus. An individual insect contains many morphologically identifiable skeletal muscles. The large number of muscles is a consequence of the segmental organization of insects and the serial replication of parts associated with segmentation. Each of the wing-bearing segments of a cockroach contains about 50 separate muscles, an abdominal segment a somewhat smaller number. In a classic anatomical study, Lyonet, in 1762, noted that the larva of the goat moth, Cossus, contains three times the number of anatomically distinct skeletal muscles as does a human. In most insect muscles the fibers lie parallel to one another, and when the muscle contracts, it shortens along the long axis of the fiber bundle. Such muscles are spoken of as being parallel-fibered muscles. In some muscles, in particular peripheral leg muscles, the fibers attach obliquely at one of their ends onto an internal, cuticular extension called an apodeme. When these muscles are activated, the muscle as a whole shortens along the axis of the apodeme, oblique to the fiber axis. The oblique insertion of fibers onto the apodeme is remindful of the oblique junction between lateral filaments and the main shaft of a feather, hence muscles with an oblique fiber arrangement are called pinnate (L. pinna = feather). The force that a muscle can generate increases with increasing cross-sectional area. The pinnate arrangement of muscles increases the effective cross-sectional area and hence the force that the muscle can produce.

Filaments and Fibrils

Muscle shortening in insects as in other animals results from sliding movement between interdigitating thick and thin filaments contained within the muscle fibers. The force of contraction is a shearing force developed between these filaments. The thick filaments are made up largely of the protein myosin, the thin filaments of the protein actin. A single thick filament is composed of many individual myosin molecules and, similarly, a thin filament contains many actin molecules. Projections of the myosin molecules from the thick filaments toward the thin filaments, called cross-bridges, are the sites of interaction between the two and are the force generators for contraction.

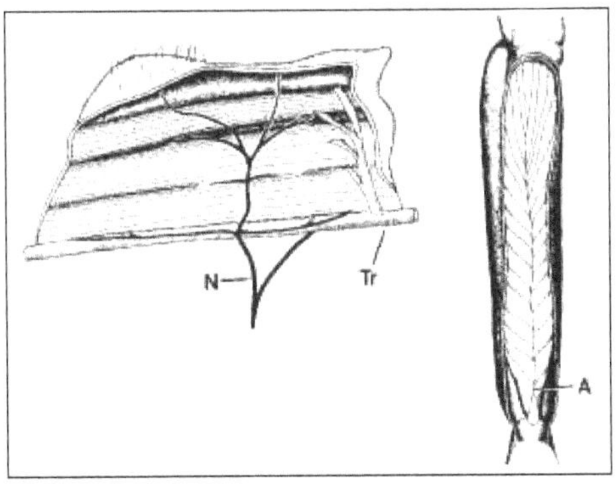

Muscle with parallel fibers (left) and one with pinnate fibers (right). The parallel-fibered muscle is the mesothoracic dorsal longitudinal flight muscle of the tettigoniid Neoconocephalus robustus. The dark structure coursing across the muscle surface is the motor nerve that innervates the muscle. The pin-nately fibered muscle is the metathoracic extensor tibia of the cricket Teleogryllus oceanicus. Abbreviations: N, motor nerve; Tr, trachea; A, apodeme.

The thick and thin filaments are grouped into longitudinal bundles called fibrils. The filaments within a fibril are grouped precisely, both longitudinally and transversely. The thin filaments attach to and project from both sides of transverse structures called Z disks. The Z disks occur regularly along the length of the fibril. The interval from one Z disk to the next is called a sarcomere. The thick filaments lay side by side in the middle of the sarcomere. The sarcomere lengths in fibrils of fast muscles such as flight muscles are 2-4 |im; those in leg muscles, body wall muscles, and visceral muscles tend to be longer, up to 7-10 |im. The regular longitudinal arrangement of Z disks, thin filaments, and thick filaments creates a striped pattern along the length of a fibril. The most obvious components of the striped pattern are (1) the Z disks; (2) the A bands, corresponding to that part of a sarcomere containing thick filaments; and (3) the I bands, corresponding to that part of a sarcomere without thick filaments. When a muscle shortens, the thin filaments slide toward the center of the sarcomere and the I bands become shorter. Because of their transverse banding pattern, muscles in insects (and skeletal muscle in vertebrates, which have a similar organization) are described as being striated muscles. The visceral muscles of insects are similar in function to vertebrate smooth muscles and in many ways similar in physiology as well. But although vertebrate smooth muscles lack striations, the visceral muscles of insects, like the skeletal muscles, are striated.

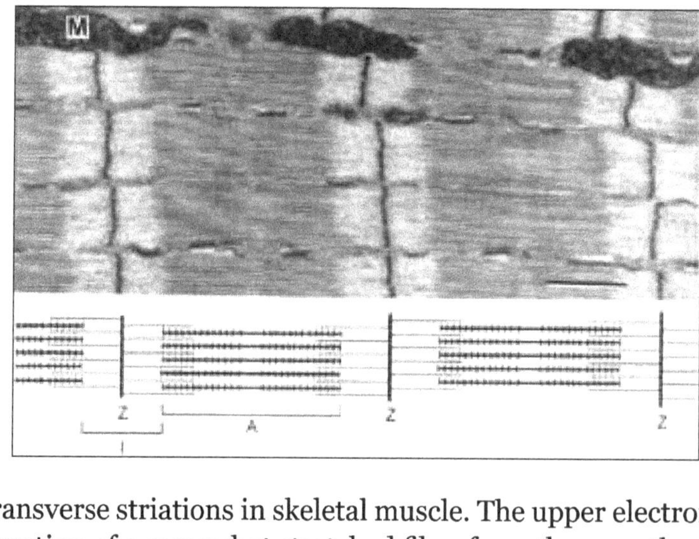

Origin of the transverse striations in skeletal muscle. The upper electron micrograph is a longitudinal section of a somewhat stretched fiber from the mesothoracic dorsal longitudinal muscle of the tettigoniid Neoconocephalus ensiger. The scale bar represents 1 |im. Abbreviations: M, mitochondrion; I, I band; Z, Z disk; A, A band.

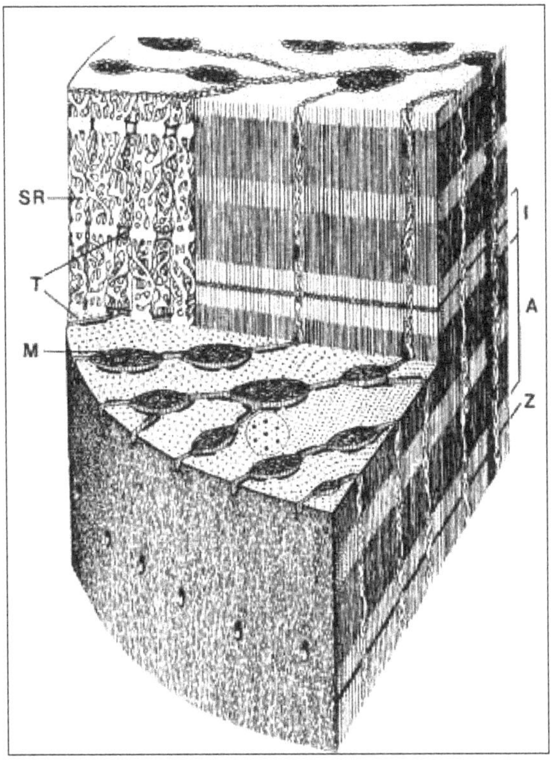

Structural organization of a fiber from an insect fast muscle. The drawing is based on electron micrographs from a tettigoniid singing muscle. The fibrils here are radial-lamellar. Abbreviations: A, A band; I, I band; M, mitochondrion; SR, sarco-plasmic reticulum; T, transverse tubule; Z, Z disk.

The thick filaments of insect muscles, and of vertebrate striated muscles, occur in a regular, hexagonal array. In vertebrate muscles a thick filament is surrounded by six thin filaments, each of which lies at the midpoint between three adjacent thick filaments, and the overall ratio of thin-to-thick filaments is 2:1. In fast muscles of insects, for example flight muscles, there are also six thin filaments surrounding each thick filament, but these occur at the midpoint between two thick filaments and the thin-to-thick ratio is 3:1. In slower insect muscles, such as body wall muscles, the usual pattern is for each thick filament to be surrounded by a circle of up to 12 thin filaments.

The fibrils of insect muscles occur in two basic patterns, cylindrical and radial-lamellar). In muscles with cylindrical fibrils the bundles of filaments forming the fibrils occur as elongate cylinders that are often polygonal in cross section. In radial-lamellar fibers the fibrils are ribbon-shaped structures arranged radially about the center of the fiber.

Other Components

The cellular components of muscle fibers seen in electron micrographs fall into four functional groups. First are those structures directly involved in the generation of force and mechanical power.

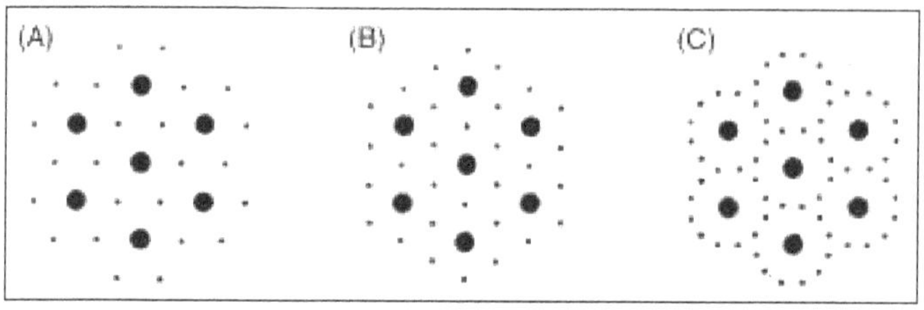

Organization of thick and thin filaments as seen in cross sections of fibers from (A) a vertebrate skeletal muscle, (B) a fast insect muscle, and (C) a slow insect muscle.

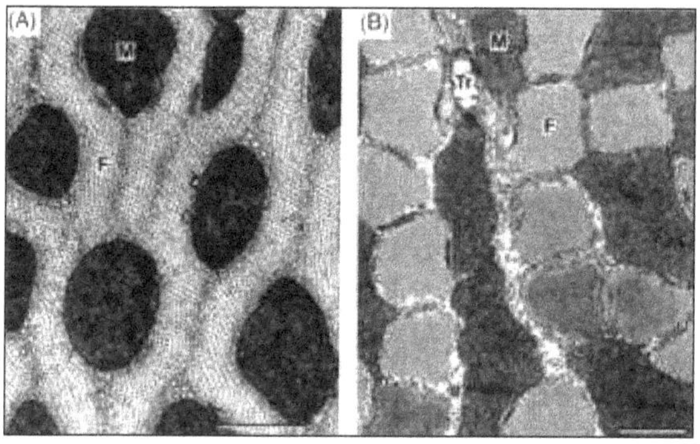

Transverse electron microscope sections through (A) a fiber with radial amellar fibrils (a flight muscle of the tettigoniid Euconocephalus nasutus) and (B) a fiber with columnar fibrils (from the tymbal muscle of the cicada Abricta curvicosta). The scale bars indicate 1. Abbreviations: M, mitochondrion; F, fibril; Tr, intrac-ellular tracheole.

These structures are the thick and thin filaments that collectively form the fibrils. Second are those components involved in the control of contraction. The most obvious structures involved in the control of contraction are the transverse tubular system (T tubules) and the sarcoplasmic reticulum (SR). The T tubules are membrane-bound tubular structures oriented perpendicular to the fiber axis. The membrane of a T tubule is continuous with the surface membrane of the fiber, and the T tubule can be regarded as an inwardly directed extension of the surface membrane. In most insect muscles there are two tubules per sarcomere, lying in the overlap areas between thick and thin filaments, but in the fibers of some muscles there is a single, centrally located T tubule per sarcomere. Within the fiber the T tubules make specialized junctions with the SR, which is an internally closed, membrane-bound compartment within the fiber. The function of the T tubules and SR is considered further below. Other elements involved in the control of contraction are the surface membrane of the muscle fibers and membrane specializations at the sites at which nerve processes contact muscle fibers. Third are the structural elements of the metabolic power supply. These are the mitochondria, which provide ATP, and glycogen granules. ATP is the immediate energy source for contraction; glycogen is a stored fuel for cellular metabolism. It would be appropriate to include among the elements involved in metabolic power the trache-oles, the terminal portions of the gas-exchange system that ramify throughout muscle fibers, even though topologically tracheoles are external to and not really part of the muscle fibers. Fourth are the structures involved in long-term maintenance of muscle, specifically the many nuclei of the fibers. The relative abundance of different cellular components in muscle is tightly correlated with the functional capacity of the muscle fibers. SR and T tubules are particularly abundant in muscles that can produce brief contractions, that is, in muscles in which the contractile apparatus can be rapidly activated and inactivated. Muscles capable of sustained activity at high power output are particularly well supplied with mitochondria and tracheolar endings. Mitochondria make up 30-40% of the muscle volume in wing muscle of active fliers and in sound-producing muscles that are active continuously and at high frequency. Such muscles are often pink, because of the cytochromes in the abundant mitochondria. It should be noted that hypertrophy of mitochondria, and of T tubules and SR, is at the expense of myofibrillar volume, so fast and fatigue-resistant muscles are likely to be relatively weak.

Muscle Attachments

Skeletal muscles attach to the cuticle of the exoskeleton through specialized epidermal cells. The muscle fibers are joined to these cells by specialized junctions called desmosomes. The terminal sarcomeres of the fibrils lack a final Z disk; instead, the thin filaments are attached to the muscle portion of the terminal desmo-some through a band of what have been called junctional filaments. Visceral muscles are frequently joined to one another by desmo-somes, and cardiac muscle fibers are joined by structures resembling the intercalary disks of vertebrate cardiac muscle.

Innervation and Activation

There is an electrical potential across the surface membrane of a living, resting muscle fiber; the interior of the fiber is typically 3070 mV electrically negative with respect to the extracellular solution. Nerve cells in the central nervous system send out long processes (motor axons) to the muscle fibers where they make specialized contacts termed synapses. A motor axon makes many synaptic contacts along the length of each muscle fiber that it innervates (multiterminal innervation), and a single muscle fiber may receive inputs from more than one motor axon (polyneuronal innervation). Impulses initiated in the central nervous system travel along the motor axons and cause the release of specific chemical signals (transmitters) from the motor axon terminals at the synapses. The transmitter released from the terminals of most motor axons leads to a reduction (depolarization) in the transmembrane potential of the muscle fiber in the vicinity of the nerve terminal. Muscle fiber depolarization initiates contraction of the fiber. Motor axons that depolarize muscle fibers and cause muscle contraction are called excitatory axons. Some axons, termed inhibitory axons, release transmitters that stabilize the transmembrane potential of the muscle fiber or even make it greater, thus antagonizing excitatory inputs. In addition to excitatory and inhibitory neural inputs, many muscles receive inputs from modu-latory motor neurons, activity that releases chemicals that modify muscle performance, for example, increasing muscle force and work output or speeding relaxation.

Insects, like other arthropods, manage their muscles using relatively few motor neurons. Some major muscles, for example tymbal muscles of cicadas, are innervated by a single motor neuron. Many muscles receive two to four motor neurons. The largest number of motor neurons yet described for an insect muscle is 16, to the flexor muscle in the leg of a locust. The processes linking membrane depolarization and contractile activation have been little studied in insect muscles, but the ultrastructure, biochemistry, and contractile performance of insect muscle are so similar to those of the far better studied frog, cat, and rodent muscles that one can predict with confidence that the basic principles worked out for vertebrate muscles apply to insects as well. The expected scheme is as follows. Membrane depolarization spreads inwardly into the fiber along the T tubules. Depolarization of the T tubules, which are coupled to the SR through specialized junctions, leads to release of calcium from the SR. Released calcium reversibly binds to regulatory sites in the fibrils, turning on the contractile machinery. Relaxation occurs as the SR takes up the released calcium and reduces the calcium concentration in the cytoplasm below that needed for contractile activity.

Muscle Mechanics

Muscle Force and Muscle Length

The muscle contraction initiated by a single stimulus, or by a single impulse in an innervating motor neuron, is termed a twitch; that evoked by repetitive input at a frequency high enough to maintain full activation of the muscle is termed a tetanus. A response in which a stimulated muscle develops force while held at constant length is called an isometric contraction. The isometric force generated by a muscle stimulated to contract in a tetanus is maximal at about the normal muscle length in the insect body and declines at longer and shorter lengths. The decline in force with increasing muscle length beyond the optimum is thought to be caused by a reduction in the overlap between the thick and the thin filaments and therefore in the number of myosin cross-bridges that can interact with the actin filaments. The decrease in force at short muscle lengths is probably a consequence of the thick filaments running into and being impeded by the Z disks, of collision of thin filaments in the middle of the sarcomere, and, at still shorter lengths, of overlap of thin filaments with portions of thick filaments of inappropriate polarity on the far side of the center of the sarcomere.

Some muscles in insects and elsewhere can shorten to a small fraction of their resting length, a response termed supercontrac-tion. The capacity for supercontraction appears to involve modifications in the structure of the Z disk such that there is not a collision between the Z disks and the thick filaments at short muscle lengths. In the supercontracting muscles that have been examined, the Z disk becomes perforate at short muscle lengths and the thick filaments slide through the spaces in the disks. The posterior, intersegmental, abdominal muscles of female locusts are of particular interest for the wide range of lengths over which they can operate. During oviposition, appendages on the end of the abdomen dig and pull the posterior abdomen down into a relatively deep hole. The intersegmental muscles become stretched to about nine times their resting length. During this stretch, called superextension, the Z disks become broken up into discontinuous, nonaligned elements to which the thin filaments are attached. Muscle contractility is not lost, and contraction of intersegmental muscles returns the abdomen to its normal length following oviposition. The latter part of the recovery may be supercontraction, for in the resting state the posterior inter-segmental muscles are normally supercontracted, with thick filaments protruding through gaps in the Z disks.

Force, Shortening Velocity and Power

There is an inverse relationship between the force on a muscle and the velocity with which it can shorten, a relationship conveniently expressed in a force-velocity plot. To facilitate comparison of muscles of differing size, force in a force-velocity plot is usually expressed as stress (force per unit cross-sectional area) and shortening velocity as strain rate (shortening velocity per unit muscle length). Two points on a force-velocity plot are frequently used to characterize a muscle's contractile properties: the maximum isometric stress of the muscle (Fmax, the intercept of the curve with the 0 velocity axis) and the maximum shortening velocity (Vmax, the intercept of the curve with the 0 force axis). Values for the maximum force in insect muscles are generally 5-35 N cm-2 (N, Newton; 1 N is approximately the downward force exerted by a mass of 100 g in the gravitational field at the earth's surface). The few available measurements of the maximum shortening velocity for insect muscles, all from fast muscles, are on the order of 3-15 lengths s. The product of force and shortening velocity has dimensions of force X distance per time (work per time) and is the rate of doing work, that is, the mechanical power output. The product of stress (force per area) and strain rate (shortening distance per second per unit muscle length) is the mechanical power per unit volume of muscle, which is readily convertible to power output per unit muscle mass. Thus each point on a force-velocity plot (or a plot of stress against strain rate) represents a power output. The power predicted from a force-velocity curve is the instantaneous power output. The peak instantaneous power is substantially greater than the sustainable power from a muscle, for during maintained activity a muscle goes through repeated contraction-relaxation cycles and therefore shortens and produces power for only part of the total time. For fast muscle, including fast insect muscles, the peak power output is 100-500Wkg-1(lW= ljs-1 = INms-1).

Muscles in insects may be divided into synchronous muscles and asynchronous muscles on the basis of the relationship between the patterns of neural activation and of contraction. The sustainable power available during repetitive, cyclic contraction has been determined for several synchronous insect muscles using the work loop approach, in which the muscle is subjected to length changes simulating those during normal activity and stimulated pha-sically during the length cycles. A plot of muscle force against muscle length for a full cycle produces a loop, the area of which is the net work output of the muscle for that cycle. The product of work per cycle and cycle frequency is the power output. The mechanical power available from synchronous flight muscles of several locusts and katydids and of a moth measured in this way ranges from 50 to 120 Wkg-1 at normal operating temperature.

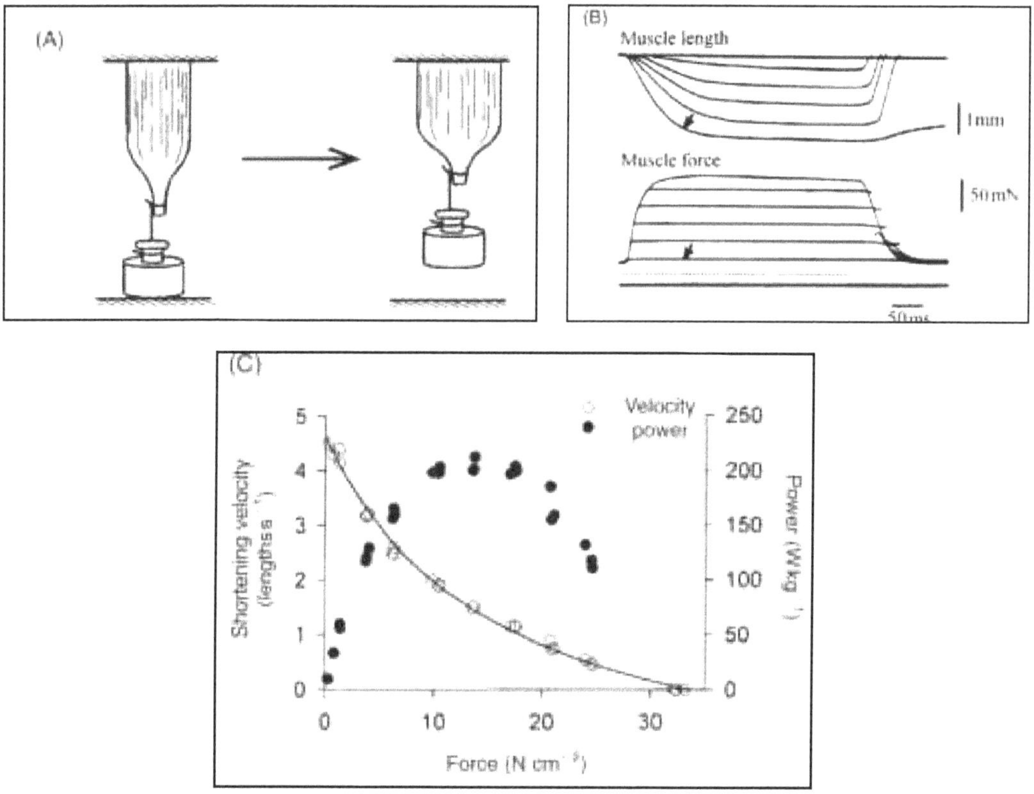

Relationships between muscle force, shortening velocity, and power output. (A) Elements of a method used to determine the relationship between force and shortening velocity. The muscle is attached to a load that is supported from below. When stimulated the muscle develops force without shortening until the force equals that of the load, following which the muscle shortens under constant load. (B) Results from an experiment examining force-velocity relations using a tettigoniid wing muscle. The lowest trace indicates the times at which the muscle was stimulated. The force trace marked by an arrow is the contraction with the smallest load of the series; the corresponding shortening trace, which has the shortest latency and the highest initial velocity, is similarly marked. (C) Force-velocity plot and a corresponding plot of power output for a wing muscle of the locust Schistocerca americana, 30 °C.

Asynchronous and Synchronous Muscles

Most insect muscles are like vertebrate skeletal muscles in that each contraction is initiated by depolarization of muscle fibers, and there is a 1:1 relationship between muscle electrical activity and muscle contraction. Such muscles may be termed synchronous muscles, reflecting the congruence between electrical and mechanical activities. The major flight muscles in several insect groups are different in that there is no synchrony between electrical and mechanical events. These muscles are known as asynchronous muscles. Neural input and fiber depolarization are needed to activate an asynchronous muscle, but when it is activated an asynchronous muscle can contract in an oscillatory manner if it is attached to a mechanically resonant load. The resonant loads for the flight muscles are the wings, which may be regarded as small, somewhat dampened tuning forks. The frequency of the oscillatory contraction is the mechanically resonant frequency of the load, which is greater than the neural input frequency required to keep the muscle fully activated. The contraction frequency of asynchronous wing muscles during flight is typically 3-10 times higher than the neural input frequency in each of the motor neurons activating the muscle. The main singing muscles in some but not all cicadas are asynchronous muscles; the resonant load here is the cuticular tymbal to which the muscle is attached and whose inward movement produces the sound pulses.

The features of asynchronous muscle that allow oscillatory contraction are stretch activation and shortening deactivation. When allowed to shorten rapidly an asynchronous muscle becomes deactivated, and while deactivated it can be stretched out to its original length, developing less force than it did while shortening. Stretching the muscle, in turn, reactivates it. Because of shortening deactiva-tion, less work is required to restretch an asynchronous muscle than is produced by the muscle during shortening, and there is net work output when the muscle undergoes a shortening-lengthening cycle. It is this net work that is available to drive the wings and power flight.

Asynchronous muscles occur in several of the most successful insect groups. They power flight in beetles, flies, bees, and wasps and many of the true bugs. The distribution of asynchronous muscles among insect taxa suggests that this mode of muscle control has evolved independently as many as 7-10 times. It is likely that asynchronous muscle has been favored by evolution because it is more powerful and more efficient than is synchronous muscle for operation at the high frequencies characteristic of insect flight. It is more powerful, in part, because asynchronous control does not demand rapidity in the rate at which muscle is turned on and off by neural input. High-frequency contraction is achieved without hypertrophy of the SR, leaving more room in muscle fibers for fibrils, which are the power-producing component. It is more efficient because a relatively low-frequency neural input is needed to maintain full activation, which reduces the amount of calcium that is released and rebound during activity and the associated metabolic costs of calcium cycling.

ENDOCRINE SYSTEM

A hormone is a chemical signal sent from cells in one part of an organism to cells in another part (or parts) of the same individual. They are often regarded as chemical messengers. Although typically produced in very small quantities, hormones may cause profound changes in their target cells. Their effect may be stimulatory or inhibitory. In some cases, a single hormone may have multiple targets and cause different effects in each target. There are at least four categories of hormone-producing cells in an insect's body:

- Endocrine glands: Secretory structures adapted exclusively for producing hormones and releasing them into the circulatory system.
- Neurohemal organs: Similar to glands, but they store their secretory product in a special chamber until stimulated to release it by a signal from the nervous system (or another hormone).
- Neurosecretory cells: Specialized nerve cells (neurons) that respond to stimulation by producing and secreting specific chemical messengers. Functionally, they serve as a link between the nervous system and the endocrine system.
- Internal organs: Hormone-producing cells are associated with numerous organs of the body, including the ovaries and testes, the fat body, and parts of the digestive system.

Together, these hormone-secreting structures form an endocrine system that helps maintain homeostasis, coordinate behavior, and regulate growth, development, and other physiological activities.

In insects, the largest and most obvious endocrine glands are found in the prothorax, just behind the head. These prothoracic glands manufacture ecdysteroids, a group of closely-related steroid hormones (including ecdysone) that stimulate synthesis of chitin and protein in epidermal cells and trigger a cascade of physiological events that culminates in molting. For this reason, the ecdysteroids are often called "molting hormones". Once an insect reaches the adult stage, its prothoracic glands atrophy (wither away) and it will never molt again.

Prothoracic glands produce and release ecdysteroids only after they have been stimulated by another chemical messenger, prothoracicotropic hormone (PTTH for short). This compound is a peptide hormone secreted by the corpora cardiaca, a pair of neurohemal organs located on the walls of the aorta just behind the brain. The corpora cardiaca release their store of PTTH only after they receive a signal from neurosecretory cells in the brain. In a sense, they act as signal amplifiers — sending out a big pulse of hormone to the body in response to a small message from the brain.

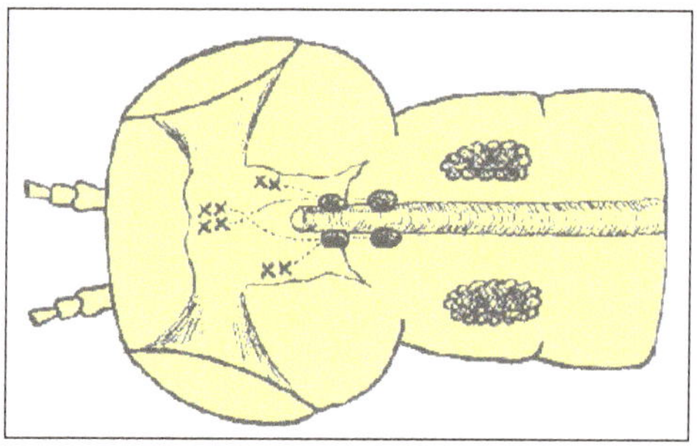

The corpora allata, another pair of neurohemal organs, lie just behind the corpora cardiaca. They manufacture juvenile hormone (JH for short), a compound that inhibits development of adult characteristics during the immature stages and promotes sexual maturity during the adult stage. Neurosecretory cells in the brain regulate activity of the corpora allata – stimulating them to produce JH during larval or nymphal instars, inhibiting them during the transition to adulthood, and reactivating them once the adult is ready for reproduction. The chemical structure of juvenile hormone is rather unusual: it is a sesquiterpene compound – more similar to defensive chemicals found in pine trees than to any other animal hormone.

The neurosecretory cells are found in clusters, both medially and laterally in the insect's brain. Axons from these cells can be traced along tiny nerves that run to the corpora cardiaca and corpora allata. The cells produce and secrete brain hormone, a low-molecular-weight peptide that appears to be the same as (or very similar to) prothoracicotropic hormone (PTTH) manufactured by the corpora cardiaca. Insect physiologists suspect that brain hormone is bound to a larger carrier protein while it is inside the neurosecretory cell, and some believe that each cluster of cells may produce as many as three different brain hormones (or hormone-carrier combinations). Large numbers of neurosecretory cells also occur in the ventral ganglia of the nerve cord, but their function is unknown.

Many other tissues and organs of the body also produce hormones. Ovaries and testes, for example, produce gonadal hormones that have been shown to coordinate courtship and mating behaviors. Ventral ganglia in the nervous system produce one compound (eclosion hormone) that helps an insect shed its old exoskeleton and another compound (bursicon) that causes hardening and tanning of the new one. There are still other hormones that control the level of sugar dissolved in the blood, adjust salt and water balance, and regulate protein metabolism.

Hormonal Control of Molting and Metamorphosis

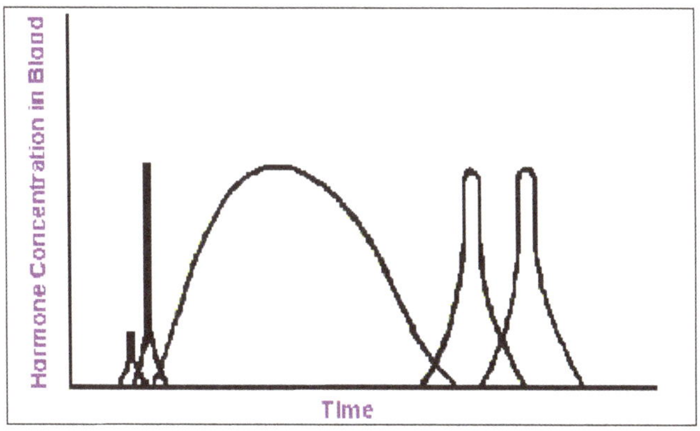

When an immature insect has grown sufficiently to require a larger exoskeleton, sensory input from the body activates certain neurosecretory cells in the brain. These neurons respond by secreting brain hormone which triggers the corpora cardiaca to release their store of prothoracicotropic hormone (PTTH) into the circulatory system. This sudden "pulse" of PTTH stimulates the prothoracic glands to secrete molting hormone (ecdysteroids).

Molting hormone affects many cells throughout the body, but its principle function is to stimulate a series of physiological events (collectively known as apolysis) that lead to synthesis of a new exoskeleton. During this process, the new exoskeleton forms as a soft, wrinkled layer underneath the hard parts (exocuticle plus epicuticle) of the old exoskeleton. The duration of apolysis ranges from days to weeks, depending on the species and its characteristic growth rate. Once new exoskeleton has formed, the insect is ready to shed what's left of its old exoskeleton. At this stage, the insect is said to be pharate, meaning that the body is covered by two layers of exoskeleton.

As long as ecdysteroid levels remain above a critical threshold in the hemolymph, other endocrine structures remain inactive (inhibited). But toward the end of apolysis, ecdysteroid concentration falls, and neurosecretory cells in the ventral ganglia begin secreting eclosion hormone. This hormone triggers ecdysis, the physical process of shedding the old exoskeleton. In addition, a rising concentration of eclosion hormone stimulates other neurosecretory cells in the ventral ganglia to secrete bursicon, a hormone that causes hardening and darkening of the integument (tanning) due to the formation of quinone cross-linkages in the exocuticle (sclerotization).

In immature insects, juvenile hormone is secreted by the corpora allata prior to each molt. This hormone inhibits the genes that promote development of adult characteristics (e.g. wings, reproductive organs, and external genitalia), causing the insect to remain "immature" (nymph or larva). The corpora allata become atrophied (shrink) during the last larval or nymphal instar and stop producing juvenile hormone. This releases inhibition on development of adult structures and causes the insect to molt into an adult (hemimetabolous) or a pupa (holometabolous).

At the approach of sexual maturity in the adult stage, brain neurosecretory cells release a brain hormone that "reactivates" the corpora allata, stimulating renewed production of juvenile hormone. In adult females, juvenile hormone stimulates production of yolk for the eggs. In adult males, it stimulates the accessory glands to produce proteins needed for seminal fluid and the case of the spermatophore. In the absence of normal juvenile hormone production, the adult remains sexually sterile.

REPRODUCTION SYSTEM

The reproductive organs of insects are similar in structure and function to those of vertebrates: A male's testes produce sperm and a female's ovaries produce eggs (ova).

Both types of gametes are haploid and unicellular, but eggs are usually much larger in volume than sperm.

Most (but not all) insect species are bisexual and biparental — meaning that one egg from a female and one sperm from a male fuse (syngamy) to produce a diploid zygote. There are, however, some species that are able to reproduce by parthenogenesis, a form of asexual reproduction in which new individuals develop from an unfertilized egg (virgin birth). Some of these species alternate between sexual and asexual reproduction (not all generations produce males), while others are exclusively parthenogenetic (no males ever occur).

Sexual reproduction might well be the most important "adaptation" ever acquired by living organisms. It provides a mechanism for shuffling and recombining genetic information from two parents to create new ("hybrid") genotypes that can be tested in the fire of natural selection. Only phenotypes that withstand the "heat" can participate in the next round of reproduction.

External vs. Internal Fertilization

Stalked spermatophore of a collembolan.

As long as primitive arthropods lived in the water, their sperm could simply swim from the male's body to the female's body where fertilization could occur. But in order to adopt a terrestrial lifestyle, animals that engaged in such external fertilization had to protect their sperm from desiccation. The solution, still used today by myriapods and insects, was to encapsulate large numbers of sperm within a water-tight lipoprotein shell secreted by the male's accessory glands. These "packages" of sperm are known as spermatophores. In myriapods and primitive hexapods (e.g. Collembola), males leave spermatophores on the ground where they may be found and picked up by a passing female. Silverfish and bristletails have more elaborate courtship activities in which the male leads his mate to a freshly deposited spermatophore.

Today, all of the more "advanced" insects exhibit internal fertilization — males deposit their sperm inside a female's body during an act of copulation. This novel adaptation, which appeared soon after insects diverged from their myriapod-like ancestors, presumably ensured that more sperm found their way to a receptive female. But the genetic programming for spermatophore production still persists in most modern insects. After a male deposits his spermatophore inside a female's reproductive system, she digests the lipo-protein coat and uses it as a source of additional nutrition for her eggs. In some cases, the quality (or quantity) of this nuptial gift may even determine whether a female accepts or rejects the male's gametes.

Sex Determination

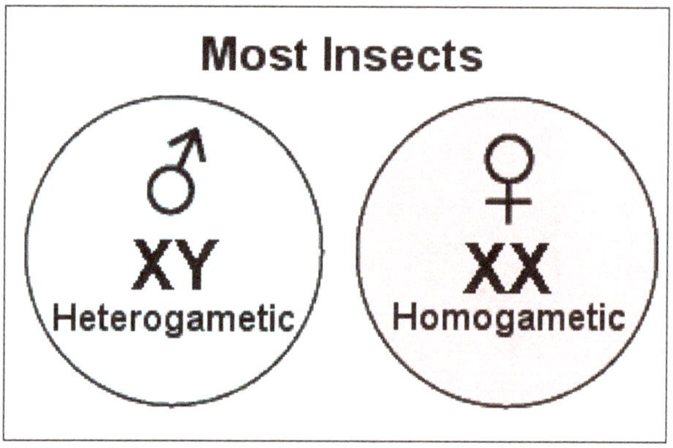

Like humans, most insects have a single pair of chromosomes that carry the genetic information for determining an individual's gender. If an embryo inherits a pair of "X" chromosomes, it will develop as a female; if it inherits one "X" and one "Y", it will develop as a male. The "XX" female is said to be homogametic; the "XY" male is heterogametic. In this case (as in humans) the male's contribution determines the offspring's gender. Some insect species have no "Y" chromosome at all — males have just one "X", and females have two. A similar condition is found in some parthenogenetic species of aphids in which "maleness" occurs through the loss (degeneration) of one chromosome during embryogenesis. In both cases, the males end up with an odd number of chromosomes (2n-1).

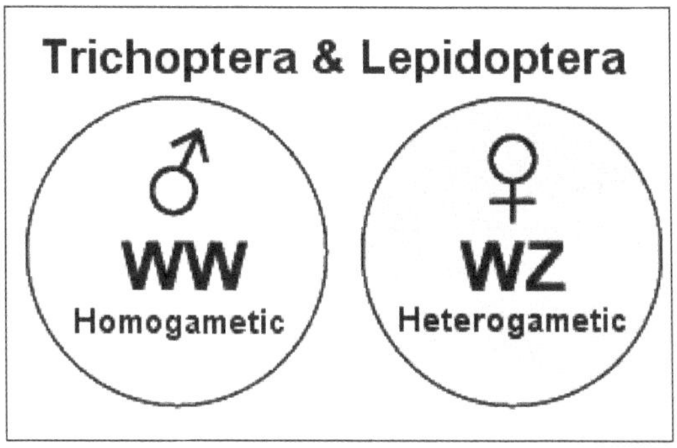

In Lepidoptera and Trichoptera, however, the homo- and heterogametic sexes are reversed: females are heterogametic and males are homogametic. To distinguish this system from standard X-Y sex determination, these sex chromosomes are designated "W" and "Z" (instead of "X" and "Y"). Thus, a female butterfly is "WZ" and a male butterfly is "WW". In this case, the female's contribution determines the offspring's gender.

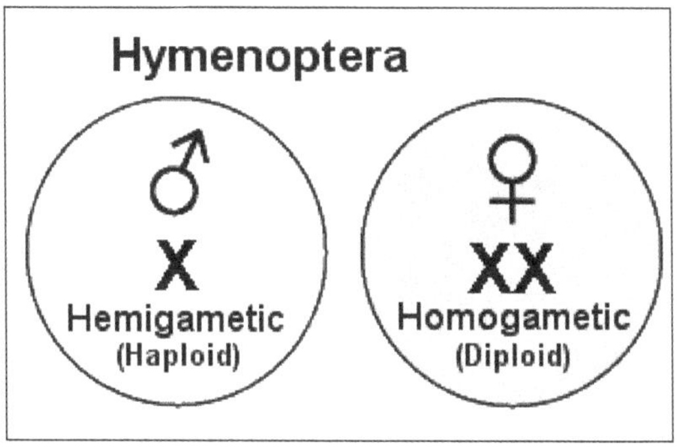

A third method of sex determination, called haplo-diploidy, is found in all Hymenoptera, many Thysanoptera, some scale insects (Hemiptera/Homoptera), and a few weevils (Coleoptera). These insects have diploid, homogametic females ("XX"), but all of the males are haploid – they develop by parthenogenesis (asexually) from unfertilized eggs. Primary oocytes undergo meiosis to form haploid eggs, but meiosis is unnecessary in primary spermatocytes because the cells are already haploid. Unmated females can lay eggs that will develop into males. Once a female mates and receives sperm from a male, she has two options:

- She can produce a female offspring by opening the valve at the base of her spermatheca to release sperm onto the egg as it passes through her oviduct.
- She can produce a male offspring by closing the spermathecal valve and preventing any sperm from reaching the egg.

Control over the gender of offspring has proven to be a useful adaptation for some insects. A biased sex ratio that favors females over males can reduce competition for limited food resources and increase the reproductive potential of the population. Bees, wasps, and ants form large colonies of queens and workers (all female) in which males are produced only sporadically as needed for reproduction.

Female Insect Reproductive System

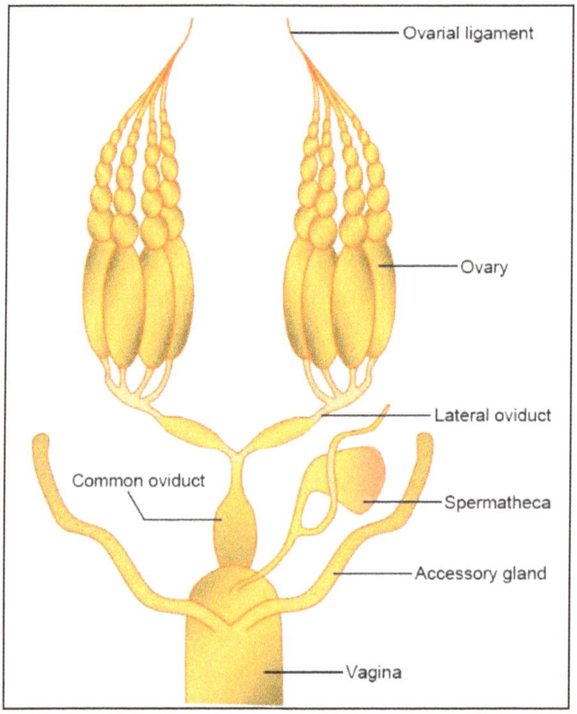

A common oviduct which enters a tubular chamber or vagina which opens to the outside via the female gonopore (genital opening) on the underside of abdominal segment. In the vast majority of insects the male transfers sperm directly into the female reproductive system, with the spermatozoa either free in fluid or encased in a capsule (of various chemical composition) to form a sperm packet or spermatophore. The sperm are stored by the female in the spermatheca sac, for later use: once she has mated she has enough sperm for life.

The spermatozoa may be deposited in the vagina (and then moved into the spermatheca) or directly into the spermatheca. In moths, butterflies and some other insects the system differs - the female has two genital openings, the opening of would have been the vagina is the ovipore and is used only for egg-laying, whilst sperm are deposited into a separate copulatory opening, which leads into a sac called the bursa copulatrix, which receives the spermatozoa before they are transferred to the spermatheca via a connecting duct called the seminal duct (which may connect to the spermatheca indirectly via an ante-chamber called the vestibule).

A female insect will typically mate with 2 or 3 males at most, and the sperm from the most recent male to mate are used preferentially. Mature eggs are passed down the oviduct and vagina from the ovaries by peristalsis (waves of contraction). As they enter the vagina the spermatheca releases some sperm to fertilise the egg. One or two pairs of accessory glands open into the vagina. The secretions from these glands include cement that coats the egg, making it sticky to enable it to be attached to a surface when deposited outside. Instead, in aquatic insects a gelatinous coat may be deposited around the egg. In the cockroach the eggs are temporarily retained in the lower vagina, whilst the cement (which contains calcium oxalate) cements a number of eggs together into a bundle. The accessory glands then secrete materials which form the egg case (ootheca) - the left accessory gland secretes protein, which coats the egg bundle, and the right secretes quinone which tans and hardens the protein on contact with air, forming the hard brown egg capsule. You may have seen female cockroaches carrying an egg capsule attached to their rear end. Some cockroaches deposit these capsules at random, others plant them with more care.

Each ovary consists of from one (some aphids) to over 2000 tubes (some termite queens)called ovarioles. The wall of each ovariole consists of a layer of epithelial cells, resting on a basement membrane, surrounded by connective tissue. At the top each ovariole tapers into a terminal filament (the terminal filaments of the ovarioles collectively forming the ovarial ligament which attaches to the body wall). Then there is a thicker elongated region called the germarium, which contains the primordial germ cells or oogonia each of which matures into either a haploid oocyte (egg cell, haploid meaning it contains half the normal number of chromosomes following a reduction division (meiosis) of the oogonium) or a nurse cell (a diploid germline cell formed by mitosis of the oogonium). The main body of the ovariole is the vitellarium, consisting of a chain of maturing oocytes, each embedded in a case of epithelial cells (normal somatic diploid cells), forming a follicle around the developing oocyte. As the oocytes mature they descend along the ovariole, so that the mature follicles are at the oviduct end and the less mature at the germarium end. Each bulge in the ovariole contains one follicle, and you can see that the more mature (older) follicles are larger at the oviduct end of the ovariole. The whole ovariole really is a chain-production line for manufacturing eggs. Finally, each ovariole is connected to the oviduct by a small ovariole stalk. When the oldest follicle is still not ready, access to this stalk is plugged by epithelial cells at the end of the vitellarium, but this plug ruptures at ovulation. The mature oocyte escapes from the ruptured follicle that encased and nourished it. This follicle then degenerates, making way for the next follicle.

There are three types of ovariole, found in different species: the first type contains no special nurse cells, the follicular cells nourish the developing egg on their own; the second and third types produce special nutritive nurse cells from the oogonia, as well as oocytes. In the second type, these nurse cells become interspersed in the follicle. In the third type they remain in the germarium and send down nutritive cords to the developing oocytes.

Male Insect Reproductive System

The male contains one pair of testes. Each testis consists of a number of tubular follicles, each consisting of a wall of connective tissue sheath, which is often pigmented. In butterflies and moths the two testes are bound together in a single connective tissue capsule.

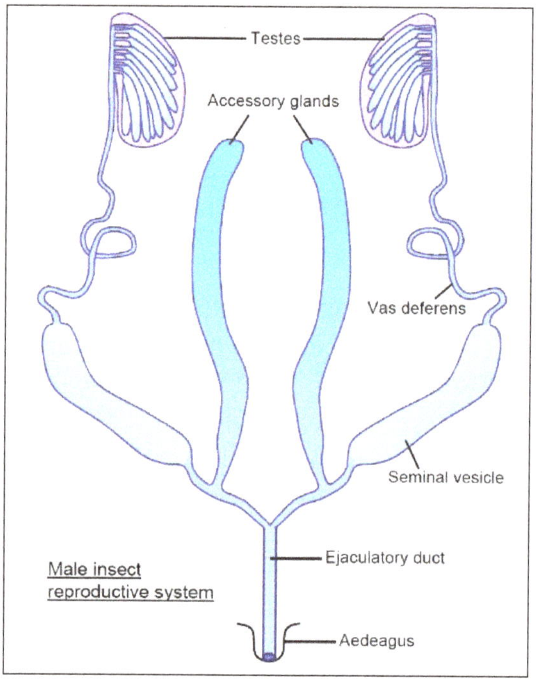

Spermatozoan production. As in the female system, each tube is a chain-production line, but making sperm cells in this case. The tip of each follicle contains densely packed primary germ cells or spermatgonia, forming a region called the germarium (or zone of spermatogonia). Large nurse cells may also be present, which nourish the developing spermatogonia. Further along the follicle is the zone of spermatocytes, are capsules of cells, called cysts, each containing 64, 128 or 256 spermatocytes, all derived from a single spermatogonium by cell division. Next follows the zone of maturation and reduction, in which each spermatocyte divides by meiosis (a reduction division in which the number of chromosomes is halved) into 4 spermatids. Then follows the zone of maturation, in which the spermatids, still encased in the cysts, develop into spermatozoa. Each grows a tail or flagellum by which the spermatozoan can swim. The writhing of these flagella eventually rupture the cyst wall and the spermatozoa escape into the vas deferens, via the stalk which connects each follicle to the vas deferens (this stalk is called the vas efferens which can be seen in the diagram, but is not labeled). The spermatozoa are often still in bundles (spermatodesms) where each bundle derived from one spermatocyte, held together by a gelatinous hyaline cap or a cyst cell (which may nourish the spermatozoa) into which the heads of the spermatozoa are embedded. In some insects as many as 16 spermatozoa are bundled together (not sure how it goes from the expected 4 to 16). The sperm become activated in the vas deferens, but they remain bundled until they enter the female, where the caps dissolve. The spermatozoa are stored in the seminal vesicles.

To summarise the sequence of sperm cell development is:

Spermatogonium (2n) - spermatocyte (2n) - spermatid (n) - spermatozoan (n).

Where 2n indicates a diploid cell (containing the normal double set of paired chromosomes) and n a haploid cell (containing only one set of chromosomes).

The seminal vesicles open into short ducts that unite to form a single ejaculatory duct, which is lined by cuticle and has powerful muscular walls. The end of this ducts is often enclosed in an evagination of the body wall, forming an intromittent organ (penis or aedeagus). A number of various accessory glands open into the ducts leading from the seminal vesicles (which may themselves secrete various substances from the duct walls). These may secrete chemicals which form the spermatophore capsule, in those insects that produce spermatophores.

References

- Digestive-system, bug-bytes: genent.cals.ncsu.edu, Retrieved 17 May, 2020
- Insect-respiration, BioTech: cronodon.com, Retrieved 05 June, 2020
- Muscle-system-insects, insects: what-when-how.com, Retrieved 02 August, 2020
- Reproductive-system, bug-bytes: genent.cals.ncsu.edu, Retrieved 18 July, 2020
- Insect-Reproduction, BioTech: cronodon.com, Retrieved 29 march, 2020
- Gullan, P.J.; P.S. Cranston (2005). The Insects: An Outline of Entomology (3 ed.). Oxford: Blackwell Publishing. pp. 61–65. ISBN 1-4051-1113-5
- Woolf CJ, Ma Q (August 2007). "Nociceptors—noxious stimulus detectors". Neuron. 55 (3): 353–64. doi:10.1016/j.neuron.2007.07.016. PMID 17678850

Evolution of Insects

CHAPTER 5

Evolution of color vision, evolution relationships, evolution of circadian clocks, evolution of eusocial insects, evolution of osmoregulatory capacity, evolution of tracheal systems, etc. are studied under the domain of evolution of insects. This chapter has been carefully written to provide an easy understanding of the subject of evolution of insects.

Insects comprise the more diverse of two classes united together as the arthropod subphylum Hexapoda, the other being the Entognatha, consisting of the orders Diplura, Protura, and Collembola (springtails). While it is often easy to recognize an insect and even a hexapod, identifying the closest relatives of Hexapoda has been a pernicious problem. Interestingly, while much has improved regarding arthropod phylogeny and the placement of hexapods, today we are somewhat less certain of a precise culprit for the hexapodan sister group. This uncertainty highlights the challenges in reconstructing relationships among major groups of Arthropoda and of interpreting broad patterns in the evolution of the phylum.

Hexapoda and the Origin of Insects

Relationships among Arthropoda have long been a matter of debate, and even monophyly of the phylum was once called into question. The door to arthropod polyphyly has been closed, however, and with the recognition of the relationship of Cycloneuralia to the panarthropod phyla (Arthropoda, Tardigrada, Onychophora), much of what we understand about arthropod evolution has been transformed. The quixotic Cambrian Rhinochelata (Opabiniida) and Radiodonta (Anomalocarida), diverging from a grade of basal lobopods, were stem groups to the Euarthropoda and its more familiar extant lineages – Chelicerata, Crustacea, Myriapoda, and Hexapoda – the latter three united as the Mandibulata and set in opposition to the chelicerates, trilobites, and trilobite-like groups.

Traditionally, Hexapoda were believed related to Myriapoda (centipedes, millipedes, and their kin), united by the presence of tracheae, or a network of exoskeletal invaginations for the transport of air and related to their shared terrestrial life. Current evidence refutes this association and considers the tracheates as independent groups at opposing ends of the mandibulate spectrum, with hexapods nested in a paraphyletic Crustacea. Many studies implicate the small, blind, and anchialine cave-inhabiting Remipedia as the living sister group to hexapods, or the more inclusive Xenocarida

(Remipedia + the benthic Cephalocarida). Morphological support for the Anartiopoda (xenocarids + hexapods) is scant, although at least among the cephalocarids the reduced number of segments and loss of abdominal appendages is somewhat hexapod-like. While Tracheata are considered defunct, support remains for such a grouping, particularly in the arrangement of pleural sclerites, the remnants of the subcoxa which are homologous to the crustacean coxa. The Xenocarida + Tracheata is tantalizing as xenocarids resemble what one would expect of a marine stem-tracheate, obviating the need to explain apparent convergences between myriapods and hexapods. Tragically, fossils of stem-group hexapods remain elusive.

Despite the challenges of identifying their nearest relatives, hexapod monophyly is strongly supported by diverse data sources. Noteworthy morphological features include the reduction of abdominal segments and appendages, legs composed of six podites, and, of course, the three appendage-bearing thoracic segments, from which is derived their name. Recognition of Entognatha is sometimes contested, with Diplura putatively closer to insects owing to the presence of cerci, paired claws, and a similar gonopore position. However, all of these traits are likely plesiomorphic (primitive), particularly if the caudal rami of xenocarids are homologous to cerci. Regardless of entognathan monophyly, the first insects were fully terrestrial, as was the common ancestor of Hexapoda, and fed on sporangia or scavenged.

Considerable effort has been expended to resolve relationships among insects, focusing almost exclusively on the modern diversity. Although there are numerous particulars that are debated, some broad patterns are consistent across sources of data and methods of analysis. No serious challenge has ever been mounted to insectan monophyly. Among the more notable of morphological characters supporting Insecta are the medial caudal filament of the eleventh abdominal segment (secondarily lost in Metapterygota), the presence of a chordotonal organ and loss of intrinsic musculature in the antenna, the loss of articulations between the thoracic sterna and coxae, and the presence of an ovipositor. Primitive insects were wingless, and the apterous orders of bristletails (Archaeognatha) and silverfish (Zygentoma) give us our closest concept of what the original insect might have resembled. These orders form a grade leading to the winged insects (Pterygota), with Zygentoma sharing with pterygotes important features in the mandible and ovipositor. Of particular interest is their shared dicondylic mandible, with its two points of articulation, which imposes unidirectional movement and permits greater force. The significance of this grouping, the Dicondylia, cannot be underestimated as it is of paramount importance for understanding one of the most significant events in Animal evolution — the origin of flight.

Insects Take to the Skies

Pterygota, the winged insects, are indisputably monophyletic, representing a single origin of wings and flight among insects. In a world where our attention is snared by charismatic birds, bats, or long-lost pterosaurs, it is easily overlooked that insects were the first animals to evolve flight and dominated the planet's skies up to 170 million years before any contenders. Flight is ancient and insects took to the air not long after arthropods invaded land, with pterygote remains dating back 410 million years. With such dramatic diversity among insects, it is no wonder that the wings themselves have been modified into any number of forms, frequently associated with the specific mechanics of flight involved, or have been coopted for purposes other than flying —defense (inclusive of crypsis), communication, mating, or thermoregulation.

It is between the silverfish-like common ancestor of Dicondylia and Pterygota that the first flyer appeared. While the functional morphology of the insect wing is robustly understood, the more abominable mystery has been the origin of the structure itself. Whereas the vertebrate wing is invariably a modified forelimb, the air foil of insects is of other derivation and pterygotes retain the full complement of six legs. Hypotheses for the origin of the insectan wing are as diverse as the organisms themselves, most based on elaborate adaptive or functional scenarios divorced from phylogenetic reasoning. Presently, phylogenetic, morphological, paleontological, developmental, and genetic evidence indicates that the wing is a planar extension of the back of the exoskeleton. The genetic architecture for a moveable joint already existed for the leg and this machinery was coopted and amalgamated to regulate formation of a hinge at the wing base. Flight first evolved to access food and aid dispersal in a world in which oxygen levels were rising and as vascular plants were making inroads into otherwise barren landscapes. Insects gave the world flight, and flight gave them the world.

Early-diverging lineages of Pterygota had wings with fixed, outstretched positions, their only extant descendants being today's mayflies (Ephemeroptera), dragonflies, and damselflies (both Odonata). Subsequent modifications permitted flexion of the wings over the abdomen and gave rise to the Neoptera. This seemingly simple act is the result of a complex arrangement of minute sclerites at the base of the wing, and the action of muscles attached to one in particular — the third axillary, a characteristically Y-shaped plate that when moved collapses the posterior portions of the wing like an accordion. Flexion permits neopterans to invade tight spaces without damage to their wings, and also paves the road for specializations unrelated to flight.

The Neoptera are organized into two principle lineages, the Polyneoptera and the Eumetabola (Paraneoptera + Holometabola). Polyneoptera are a heterogeneous assemblage of orders and covers the earwigs (Dermaptera), crickets, katydids, and grasshoppers (Orthoptera), stick and leaf insects (Phasmatodea), stoneflies (Plecoptera), webspinners (Embiodea), zorapterans (Zoraptera), crawlers (Notoptera), mantises (Mantodea), roaches (Blattaria), and termites (Isoptera). The Paraneoptera includes the barklice (Psocoptera), true lice (Phthiraptera), thrips (Thysanoptera), and the prominent plant-feeding true bugs (Hemiptera). While one should never disparage the variety resident among Polyneoptera and Paraneoptera, when truly speaking of insect diversity it is in reference to Holometabola. Holometabola include those orders with complete metamorphosis, an innovation that, like flight, was crucial for insectan hegemony.

A Developmental Shift

Holometaboly, or complete metamorphosis, is typified by a soft-bodied, morphologically-reduced larva, followed by a largely quiescent pupa. The larva often has a diet and life quite independent from that of the adult, permitting a single species to effectively live divergent existences. Larvae often occur in protected habitats and can enter prolonged periods of diapause during times of stress. Archaeognatha and Zygentoma have virtually no change in appearance from nymph to the sexually mature adult, owing to the absence of wing buds (ametaboly), and molting occurs throughout life. Hemimetaboly, or incomplete metamorphosis, encompasses a range of varied developmental patterns, but shares morphological resemblance across all stages, nymphs effectively being miniaturized adults with wing buds. Fully-functional wings and sexual maturity appear in the adult at which point molting ceases. Mayflies have a transitional form whereby wings become functional one molt prior to the adult – the subimago. This subimaginal molt is homologous to the post-adulthood molts of ametabolous insects, and the loss of this stage is a dramatic feature of Metapterygota. From an evolutionary perspective, the larva appears to represent a protracted hemimetabolan pronymph. The pronymph is a brief, often-unnoticed stage between hatching and the first nymphal instar, and pronymphs sometimes never depart the egg. Pronymphs share anatomical, physiological, and embryological features with larvae, and prolongation of this stage provides greater control over development. The pupa is conversely a compaction of the various nymphal stages prior to eclosion as an adult.

The Holometabola comprises nearly 85% of insect diversity and today contains more species than there are among plants or all other animal phyla combined. Orders include the ants, bees, and other wasps (Hymenoptera), lacewings and antlions (Neuroptera), dobsonflies and alderflies (Megaloptera), snakeflies (Raphidioptera), twisted-wing parasitoids (Strepsiptera), beetles (Coleoptera), scorpionflies (Mecoptera), nannochoristids (Nannomecoptera), snow fleas (Neomecoptera), true fleas (Siphonaptera), true flies (Diptera), caddisflies (Trichoptera), and moths and butterflies (Lepidoptera). The advent of the larva did not immediately confer tremendous advantages that led to a proliferation of species, as holometabolans were spectacularly meager for epochs after their emergence. Moreover, it seems heretical but Holometabola per se are not diverse, as it is only from among apocritan wasps, staphylinoid and phytophagan beetles, higher flies, and ditrysian moths that the species-rich Behemoths emerge.

Modern hexapod diversity representing the summation of over 400 million years of high speciation rates and low levels of natural extinction.

Nature's inordinate fondness for six legs.

Each of these were independent triumphs achieved at different times. There was no explosive holometabolan or 'beetle' radiation, and we concatenate into single episodes eons of shifting climates, changing geography, and simple chance, speaking of evolutionary radiations as though they took place in ecological space and time and as adaptive responses. Yet, ecological interactions operate quickly, sometimes within dozens of generations, and lose their impact when attempting to explain patterns covering tens of millions of years. Nonetheless, the appearance of a larva permitted any number of subsequent evolutionary innovations key to their prosperity, such as endoparasitism.

FOSSIL RECORD OF INSECTS

Until recently, the oldest fossil hexapods were Collembola, including Rhyniella praecursor, known from about 400 mya (million years ago) in the Lower Devonian of Rhynie, Scotland, and slightly younger archaeognathans from North America. Reinterpretation of another Rhynie fossil, Rhyniognatha hirsti, known only from its mouthparts, suggests that it is the oldest "ectognathous" insect. Tantalizing evidence from Lower Devonian fossil plants shows damage resembling that caused by the piercing-and-sucking mouthparts of insects or mites. Any earlier fossil evidence for Insecta or their relatives will be difficult to find because appropriate freshwater fossiliferous deposits are scarce prior to the Devonian.

In the Carboniferous, an extensive radiation is evidenced by substantial Upper Carboniferous fossils. Lower Carboniferous fossils are unknown, again because of lack of freshwater deposits. By some 300 mya a probably monophyletic grouping of Palaeodictyopteroidea comprising three now-extinct ordinal groups, the Megasecoptera, Palaeodictyoptera, and Diaphanopterodea, was diverse. Palaeodictyopteroideans varied in size (with wingspans up to 56 cm), diversity (over 70 genera in 21 families are known), and in morphology, notably in mouthparts and wing articulation and venation. An apparently paraphyletic "Protodonata", perhaps a stemgroup of Odonata, had prothorax winglets as did Palaeodictyopteroidea, and included Permian insects with the largest wingspans ever recorded. Extant orders represented unambiguously by Carboniferous fossils are limited to the Orthoptera; putative Ephemeroptera, fossil hemipteroids, and blattoids are treated best as paraphyletic groups lacking the defining features of any extant clade. The beak-like, piercing mouthparts and expanded clypeus of some Carboniferous insects indicate an early origin of plant-feeding, although it was not until the Permian that gymnosperms (conifers and allies) became abundant in the previously fern-dominated flora. Concurrently, a dramatic increase took place in ordinal diversity, with some 30 orders known from the Permian. The evolution of the plant-sucking Hemiptera may have been associated with the newly available plants with thin cortex and sub-cortical phloem. Other Permian insects included those that fed on pollen, another resource of previously restricted supply.

Certain Carboniferous and Permian insects were very large, exemplified by the giant Bolsover dragonfly and palaeodictyopteran on a *Psaronius* tree fern, depicted in the vignette for this chapter. Fossils of some Ephemeroptera and dragonfly-like Protodonata had wingspans of up to 45 cm and 71 cm respectively. A plausible explanation for this gigantism is that the respiratory restriction on insect size might have been alleviated by elevated atmospheric oxygen levels during the late Palaeozoic (between 370 and 250 mya), which would have promoted greater oxygen diffusion in the tracheae. Furthermore, if other gases were unchanged, then any extra atmospheric oxygen may have facilitated flight in denser air. Despite the obvious appeal of this hypothesis to account for Palaeozoic gigantism, such a climatic reconstruction remains controversial and the physiological and morphological consequences of alterations of gaseous composition have been little studied for insects.

Many groups present in the Permian, including Ephemeroptera, Odonata, Plecoptera, Grylloblattodea, Phasmatodea, Orthoptera, Dictyoptera and other Polyneoptera, plus Coleoptera and Neuropterida amongst other early originating Holometabola, survived the period. However, early lineages such as Palaeodictyopteroidea (including Diaphanopteroidea, Megasecoptera, and Palaeodictyoptera) disappeared at the end of the Permian. This Permian—Triassic boundary was a time of major extinction that particularly affected marine biota, and coincided with a dramatic reduction in diversity in taxa and feeding types within surviving insect orders.

The Triassic period (commencing about 245 mya) is famed for the "dominance" by dinosaurs and pterosaurs, and the origin of the mammals; but the insects were radiating apace. The major orders of modern insects, excepting Hymenoptera and Lepidoptera, were well established by the beginning of the Triassic. Hymenoptera are seen first in this period, but represented only by symphytans. The oldest living families appeared, together with diversified taxa with aquatic immature (and some adult) stages, including modern Odonata, Heteroptera, and many families of nematocerous Diptera. The Jurassic saw the first appearance of aculeate Hymenoptera, many early-branching Diptera, and the first Brachycera. Triassic and Jurassic fossils include some excellent preserved material in fine- grained deposits such as those of Solenhofen, the site of beautifully preserved insects and *Archaeopteryx*. The origin of birds (Aves) and their subsequent diversification marked the first aerial competition for insects since the evolution of flight.

In the Cretaceous (145–65 mya) and throughout the subsequent Tertiary period (65–1.6 mya), excellent arthropod specimens were preserved in amber – a resinous plant secretion that trapped insects and hardened into a clear preservative. The excellence of preservation of whole insects in amber contrasts favorably with compression fossils that may comprise little more than crumpled wings. Many early fossil records of extant higher taxa (groups above species level) derive from these well-preserved amber specimens, but inherent sampling biases must be recognized. Smaller (more easily trapped) and forest-dwelling taxa are over-represented. Amber of Cretaceous origin occurs in France, Lebanon, Burma, Siberia, Canada, and New Jersey. The biota of this period shows a numerical dominance of insects coincident with angiosperm (flowering plant) diversification. However, major mouthpart types of extant insects evolved prior to the angiosperm radiation, associated with insect feeding on earlier terrestrial plant radiations. The fossil record indicates the great antiquity of certain insect–plant associations. The lower Cretaceous of China (130 mya) has revealed both early angiosperms and a distinctive fly belonging to Nemestrinidae with a characteristic long proboscis associated with angiosperm pollination. Elsewhere, a fossil leaf of an ancestral sycamore has the highly characteristic mine of the extant genus *Ectoedemia* (Lepidoptera: Nepticulidae), suggesting at least a 97 million year association between the nepticulid moth and particular plants. Both Coleoptera and Lepidoptera, which are primarily phytophagous orders, commenced their massive radiations in the Cretaceous. By 65 mya, the insect fauna looks rather modern, with some fossils able to be allocated to extant genera. For many animals, notably the dinosaurs, the Cretaceous–Tertiary ("K–T") boundary marked a major extinction event. Although it is generally believed that the insects entered the Tertiary with little extinction, recent studies show that although generalized insect–plant interactions survived, the prior high diversity of specialist insect–plant associations was greatly attenuated. At least in the paleobiota of south-western North Dakota, at 65 mya a major ecological perturbation setback specialized insect–plant associations.

Our understanding of Tertiary insects increasingly comes from amber from the Dominican Republic dated to the Oligocene/Miocene (37–22 mya) to complement the abundant and well-studied Baltic amber that derives from Eocene/Oligocene deposits (50–30 mya). Baltic ambers have been preserved and are now partially exposed beneath the northern European Baltic Sea and, to a lesser extent, the southern North Sea, brought to shore by periodic storms. Many attempts have been made to extract, amplify, and sequence ancient DNA from fossil insects preserved in amber – an idea popularized by the movie *Jurassic Park*. Amber resin is argued to dehydrate specimens and thus protect their DNA from bacterial degradation. Success in sequencing ancient DNA has been claimed for a variety of amber-preserved insects, including a termite (30 myo (million years old)), a stingless bee (25–40 myo), and a weevil (120–135 myo); however, attempts to authenticate these ancient sequences by repetition have failed. Degradation of DNA from amber fossils and contamination by fresh DNA are unresolved issues, but the extraction of material from amber insects leads to certain destruction of the specimen. Even if authentic DNA sequences have been obtained from amber insects, as yet they do not provide more useful information than that which derives from the often exquisitely preserved morphology.

An unchallengeable outcome of recent studies of fossil insects is that many species and higher taxa, especially genera and families, are revealed as much older than thought previously. At species level, all northern temperate, sub-Arctic, and Arctic zone fossil insects dating from the last million years or so appear to be morphologically identical to existing species. Many of these fossils belong to beetles (particularly their elytra), but the situation seems no different amongst other insects. Pleistocene climatic fluctuations (glacial and interglacial cycles) evidently caused taxon range changes, via movements and extinctions of individuals, but resulted in the genesis of few, if any, new species, at least as defined on their morphology. The implication is that if species of insect are typically greater than a million years old, then insect higher taxa such as genera and families may be of immense age.

Modern microscopic paleontological techniques can allow inference of age for insect taxa based on recognition of specific types of feeding damage caused to plants which are fossilized. As seen above, leaf mining in ancient sycamore leaves is attributable to a genus of extant nepticulid moth, despite the absence of any preserved remains of the actual insect. In like manner, a hispine beetle (Chrysomelidae) causing a unique type of grazing damage on young leaves of ginger plants (Zingiberaceae) has never been seen preserved as contemporary with the leaf fossils. The characteristic damage caused by their leaf chewing is recognizable, however, in the Late Cretaceous deposits from Wyoming, some 20 million years before any body fossil of the culprit hispine. To this day, these beetles specialize in feeding on young leaves of gingers and heliconias of the modern tropics.

Despite these valuable contributions made by fossils, several tempting inferences concerning their use in phylogenetic reconstruction ought to be resisted, namely that:

- All character states in the fossil are in the ancestral condition.
- Any fossil represents an actual ancestral form of later taxa.
- The oldest fossil of a group necessarily represents the phylogenetically earliest taxon.

Notwithstanding this last point, fossil insects tend to demonstrate that the stratigraphic (time) sequence of earliest-dated fossils reflects the branching sequence of the phylogeny. Early-branching (so-called "basal") taxa within each order are represented first — for example, the Mastotermitidae before the higher termites, midges, gnats, and sand flies before house and blow flies, and primitive moths before the butterflies. Furthermore, insect fossils can show that taxa currently restricted (or perhaps disjunct) in distribution once were distributed more widely. Some such taxa include the following:

- Mastotermitidae (Isoptera), now represented by one northern Australian species, is diverse and abundant in ambers from the Dominican Republic (Caribbean), Brazil, Mexico, USA, France, Germany, and Poland dating from Cenomanian Cretaceous (c. 100 mya) to lower Miocene (some 20 mya).
- The biting-midge subfamily Austroconopinae (Diptera: Ceratopogonidae), now restricted to one extant species of *Austroconops* in Western Australia, is represented by several species in Lower Cretaceous Lebanese amber (Neocomian, 120 mya) and Upper Cretaceous Siberian amber (90 mya).
- *Leptomyrmex*, an ant genus now distributed in the western Pacific (eastern Australia, New Guinea, New Caledonia), is known from 30–40 myo Dominican amber.

An emerging pattern stemming from ongoing study of amber insects, especially those dating from the Cretaceous, is the former presence in the north of groups now restricted to the south. We might infer that the modern distributions, often involving Australia, are relictual due to differential extinction in the north. The question of whether such patterns relate in any way to northern extinction at the K—T boundary due to bolide ("meteorite") impact deserves further study. Did the differential extinction of specialist insect—plant associations in South Dakota extend to the southern hemisphere? Evidently, some insect taxa presently restricted to the southern hemisphere but reported as having been present 30–40mya in Dominican and subsequent Baltic ambers did survive the event and regional extinction has occurred more recently.

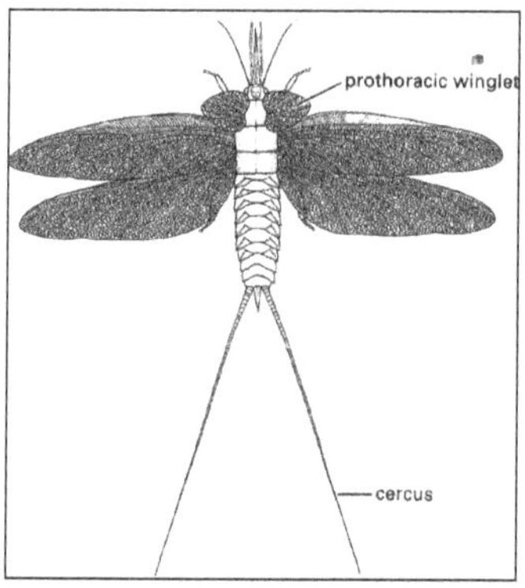

Reconstruction of Stenodictya lobata (Palaeodictyoptera: Dictyoneurida.

The relationship of fossil insect data to phylogeny derivation is complex. Although early fossil taxa seem often to precede phylogenetically later-branching ("more derived") taxa, it is methodologically unsound to assume so. Although phylogenies can be reconstructed from the examination of characters observed in extant material alone, fossils provide important information, not least allowing dating of the minimum age of origin of diagnostic derived characters and clades. Optimally, all data, fossil and extant, can and should be reconciled into a single estimate of evolutionary history.

e).

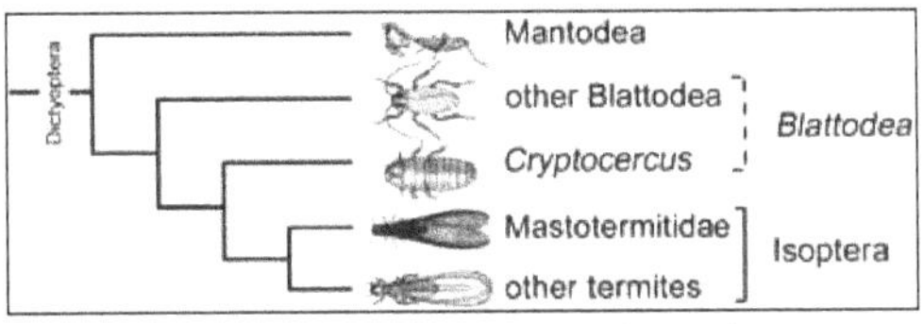

Cladogram of postulated relationships within Dictyoptera, based on combined morphological and nucleotide sequence data.

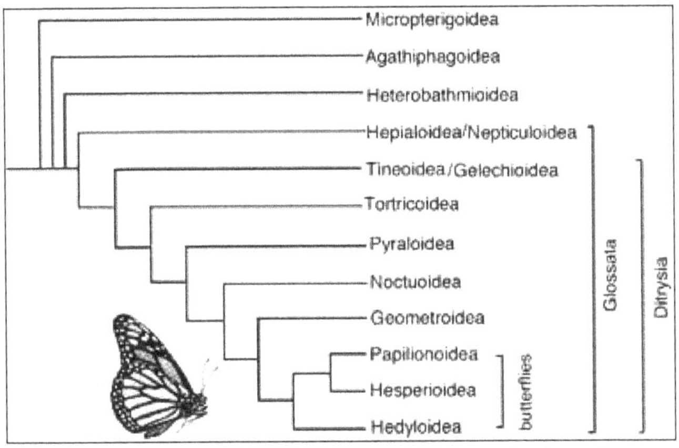

Cladogram of postulated relationships of selected lepidopteran higher taxa, based on morphological data.

Taxa that contain only fossils are indicated by the symbol †. The fossil record of Protura, Phthiraptera, and Zoraptera dates from only the Holocene, presumably because of the sma ll size and/or rarity of these organisms. Miomoptera represent the earliest fossil endopterygotes. The placement of *Rhyniognatha* is unknown.

EVOLUTIONARY PATTERNS OF OSMOREGULATORY CAPACITY IN INSECTS

Two main osmoregulatory adaptations have been identified that permit insects to thrive in the terrestrial environment: a waterproofed cuticle and the capacity to produce a hyperosmotic excreta. The cuticle, which is secreted by the epidermis, contains multiple proteins and the carbohydrate chitin. These form the flexible, yet tough, exoskeleton of insects. The waterproofing aspects of the cuticle are thought to be due almost entirely to an epicuticular layer of waxes and oils secreted by the underlying dermal glands. The second adaptation to terrestrial life, the capacity to produce hyperosmotic excreta, derives in insects from the functions of the gut. The midgut of insects is the site of digestion and absorption of ingested food and fluids. The Malpighian tubules, which empty into the gut near the midgut, are the site of primary urine production. The fluids in the lumina of both the midgut and Malpighian tubules are iso-osmotic to the hemolymph. This fluid moves within the gut to the rectum. There, a single cell type is used to extract water from the gut contents in the rectal lumen, thereby transporting a hyposmotic fluid to the hemolymph. As a result of this activity, a hyperosmotic excreta is expelled via the anus.

These two adaptations, a cuticle highly impermeable to water and a rectum capable of producing hyperosmotic excreta, are found in virtually all insects lineages. It is clear that these are primitive characters that occurred early in insect evolution and were central to the success of insects on land.

Among terrestrial insects, a few species have evolved another very useful trait, namely the capacity to extract water from air. A number of species can extract water from air using the rectum. In these species, the ion-transporting capacities of the Malpighian tubules are employed, however a highly impermeable cryptonephridial barrier surrounding both the tubules and the rectum prevents water from the hemolymph from following. As ions accumulate in the cryptonephridial space, a hyperosmotic fluid is produced which draws not from the hemolymph but rather from the excreta in the rectal lumen. These same insects can pump air into and out of the rectal lumen via the anus. Under these circumstances, water can move down its activity gradient from the air in the rectum to the cryptonephridial space. From there, the water flows into the hemolymph. Other insects can remove water from subsaturated air by secreting a hyperosmotic, hygroscopic fluid onto surfaces near the mouth. After the fluid absorbs water, it is ingested. Phylogenetic analysis reveals that both of these strategies have evolved rarely and independently in terrestrial species. These capacities are not a primitive trait.

Many insect species have exploited aquatic habitats, particularly during their larval stages. The number of species and their abundance is testimony to the ecological success of insects in this habitat. Most of the aquatic insects are restricted to fresh water. The principle adaptation leading to success in fresh water involves the rectum. In freshwater insects, the rectum is the site of ion uptake from the excreta and primary urine, with little water following. The rectal cells of freshwater insects actively resorb ions, particularly sodium, chloride, and potassium. In many very oligotrophic habitats, however, the urine cannot be made as dilute as the surrounding waters. In these circumstances "extrarenal sites" of ion uptake are employed that absorb ions from the external medium, with little water following. These functions are carried out by specialized chloride cells as found, for example, on the gills of mayflies, or in the anal papillae of mosquito larvae. Phylogenetic analysis demonstrates that adaptation to freshwater habitats occurred independently several times in the insects; however, terrestrial existence was the primitive condition.

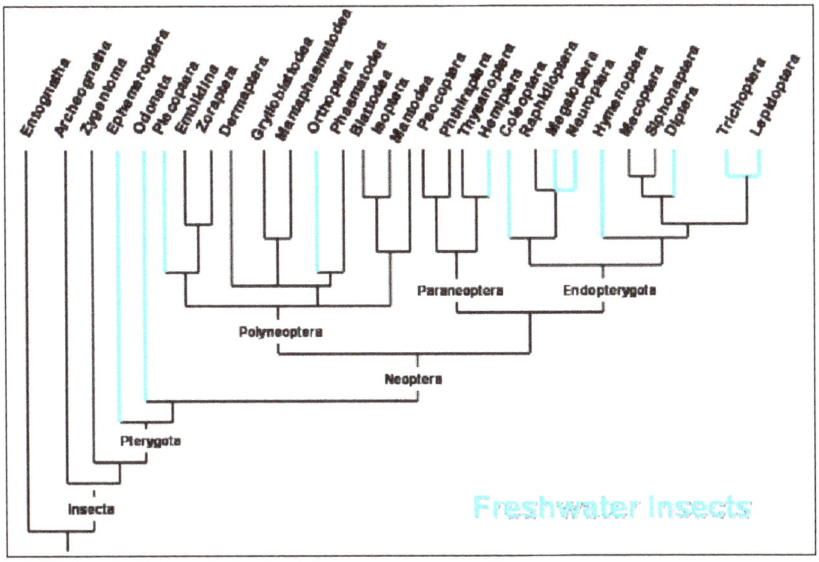

A depiction of phylogenetic relationships in the insects. The lines in blue indicate lineages in which some species can inhabit fresh water. The relationships shown would suggest that the capacity to inhabit fresh water evolved repeatedly and independently in various lineages within the insect clade.

A few insect orders contain species that can survive in saline waters. Again, the principal adaptation involves cells in the hindgut. In saline-water mosquitoes in the genus Ochlerotatus, the rectum is divided into two segments. The anterior segment serves to take up ions from the primary urine and excreta. This function is vital when the larvae find themselves in fresh water, but also is useful in saline waters as a means of recycling valuable ions from the primary urine. The posterior rectal segment functions as a salt gland. It is capable of transporting sodium, magnesium, chloride and possibly bicarbonate, as necessary to remove ions obtained by ingesting the external medium. These ions are transported with little water following, resulting in a highly saline excreta. The insects that can survive in saline waters are all in orders also containing freshwater species. It would seem that adaptation to saline water is a highly derived characteristic, probably arising in species already adapted to fresh water. Saline tolerance occurs very infrequently in the insects and is clearly not a primitive state for the group.

It appears that cuticular and rectal adaptations that support terrestrial life occurred very early in the insects and are primitive characters for the entire clade. Some terrestrial insects also evolved a capacity to take up water from a subsaturated atmosphere. Adaptation to an aquatic existence occurred much later during insect evolution, and occurred frequently and independently in several insect orders. Adaptation to saline waters seems to have originated in freshwater ancestors. The principle adaptations for aquatic insects are modifications of rectal morphology and function; in freshwater forms, "chloride cells" take up ions from the external medium.

TRACHEAL SYSTEMS AND THE EVOLUTION OF INSECTS

Most insects respire via an air-filled tracheal respiratory system. The tracheal system generally consists of cells joined to form walls of hollow tubes that run throughout the body, terminating in blind-ended tracheoles in close proximity to the cells. The use of air instead of a liquid (e.g., blood) as a medium for delivering oxygen comes with several major advantages: (1) It is light-weight and requires less energy for production, maintenance, and transport compared to blood-based systems. (2) It supports aerobic metabolism during active flight, which demands the highest mass-specific rates of oxygen consumption in the animal kingdom. The evolution of flight is a very important key to the great biodiversity and the ecological success of insects. (3) It provides an internal reservoir for oxygen, which allows prolonged periods of asphyxia (e.g., in hypoxic environments or under water). (4) If insects eventually become paralyzed by exposure to anoxia, they can recover quickly when re-exposed to oxygen, because oxygen can be delivered by passive diffusion, which is rarely true in vertebrates. However, the buoyancy associated with possession of a tracheal system may cause insects to be excluded from pelagic and deepwater environments.

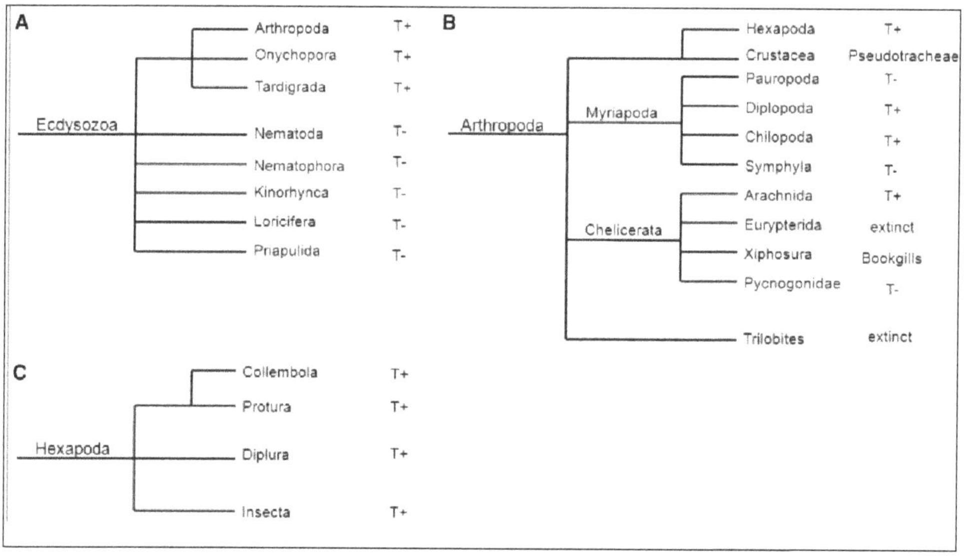

T+ represents the presence of tracheae in at least one species, T− represents no tracheae present in any species (A) Tracheal distribution across Ecdysozoa. (B) Tracheal distribution across Arthropoda. (C) Tracheal distribution across Hexapoda.

Despite the physiological importance of the tracheal system to insect function and success, the evolution (origin and diversification) of the tracheal system is very poorly understood. Tracheal systems appear in all three related clades of Ecdysozoa (animals that shed their exoskeleton in a molt) that achieve large size and terrestriality—arthropods, onychophorans, and tardigrades. Yet, all three tracheal systems would appear to have evolved independently since we know that primitive arthropods and tardigrades were aquatic. Within arthropods, all three major groups—the myriapods, chelicerates, and pancrustaceans (the taxonomic group containing hexapods and crustaceans)—exhibit tracheae. Again, there is evidence for multiple independent evolutions of the tracheal system—(1) current phylogenies have the closest ancestor to hexapods being an aquatic crustacean that lacks tracheae, a branchiopod, indicating an independent evolution of tracheae within hexapods, (2) most of the chelicerates are aquatic and lack tracheae pointing to an independent evolution of tracheae within the arachnids, and (3) if we accept a branchiopod relationship to hexapods, the myriapods, which are all terrestrial and possess tracheae, must also have an independently evolved tracheal system.

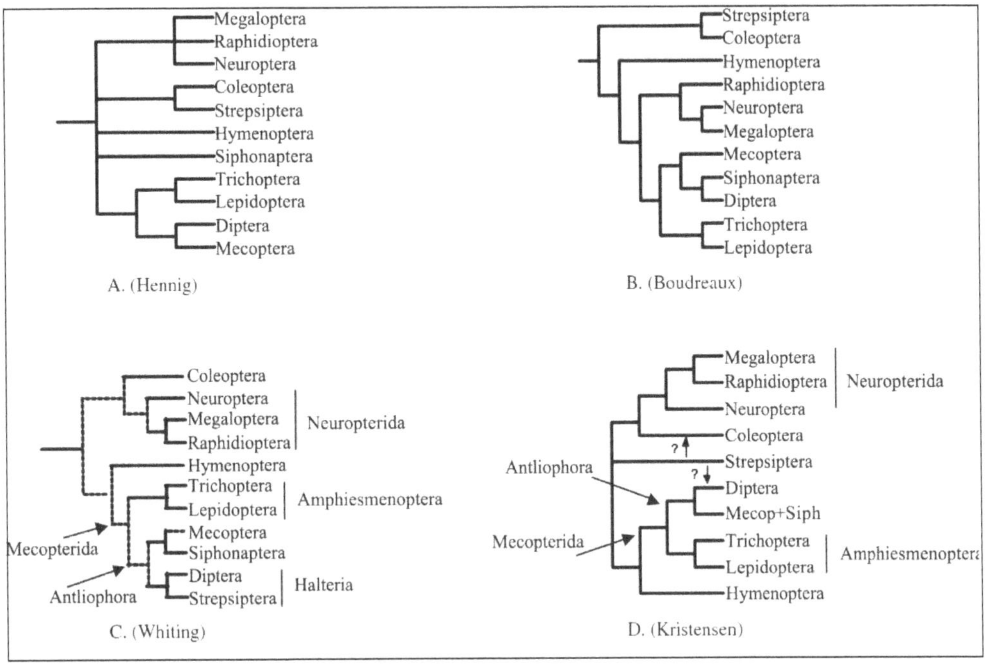

Within hexapods, the vast majority of insects possess tracheae and it has generally been assumed that these systems all derive from a common ancestor, but given the number of independent evolutionary events apparent at the basal parts of the tree and examples of secondary loss in aquatic and small forms, even this may no longer be certain. However, while the tracheae or tracheal-like structures in a vast diversity of ecdysozoan groups may not derive from a common ancestor, we posit that all Ecdysozoa possess some fundamental developmental pathways that can be utilized in the formation of tracheae. All Ecdysozoa possess a cuticle and perhaps developmental signals inducing an invagination of this cuticle may underlie tracheal formation. This hypothesis is supported by the fact that all terrestrial groups (although not all species) of Ecdysozoa exhibit tracheae. At the moment, there is only one species in which the molecular mechanisms of tracheal development have been well-studied: the fruitfly *Drosophila melanogaster*. The fates of tracheal cells are specified by the interactions of the heterodimeric transcription factor encoded by *trachealess* and *tango*, which together cause tracheal precursor cells to undergo cytoskeletal rearrangements necessary for invagination. It would be very interesting to examine the developmental pathways and genetic controls of the tracheal system in a much wider array of Ecdysozoa to discern if genes similar to *trachealess* and *tango* determine the fate of tracheal cells in the various independently evolved tracheal systems. Also, understanding the developmental role of such genes in aquatic Ecdysozoa that lack tracheae might help reveal the evolutionary pathway. Interestingly, *tango* codes for the protein that functions as the constitutive component of hypoxia-inducible factor, an oxygen sensor in all animal groups, suggesting that it might function in gills or other exoskeletal structures that might respond to oxygen levels in aquatic organisms.

On an evolutionary scale, the possession of a tracheal system may make insects more susceptible to changes in atmospheric composition through geologic time than is true for other groups. One of the most striking examples of this is the arthropod gigantism in the Carboniferous and Permian, coincident with hyperoxic atmospheres reaching 30% oxygen. During these times, there were Protodonata with 70 cm wingspans and Arthropleura that were 2 m in length. The mechanism behind this gigantism may be linked to tracheal respiratory systems.

The tracheal system limitation hypothesis suggests that as insects get larger, the blind-ended tracheal respiratory system leads to increasing challenges for gas exchange due to the effects of distance on diffusion rates. To compensate, insects increase their mass-specific investment in the tracheal system and utilize more convection. Eventually, this increased investment leads to spatial constraints that limit insect size. When animals are reared under higher levels of atmospheric oxygen, their tracheae are smaller, mass-specific investment in the tracheal system is reduced, and the insects can therefore attain larger sizes before reaching the limits set by spatial constraints. This provides one possible mechanism whereby increasing atmospheric oxygen through time could lead to insect gigantism. In addition to effects on tracheal structures, rearing extant insects under different oxygen levels has strong effects on average body size and developmental rates in a variety of insect groups, suggesting that changes in atmospheric composition in the past would have greatly influenced insect ecology and evolution. Both of these mechanisms indicate that the insects' unique physiology and possession of a tracheal system have been fundamental to their evolutionary history and the impact that environmental change has had on their evolutionary trajectory.

EVOLUTION OF COLOR VISION IN HOLOMETABOLOUS INSECTS

The eyes of insects are remarkable. Much of the diversity in eyes can be traced to alterations in the number, spectral properties, and spatial distribution of the visual pigments. Visual pigments are light-sensitive molecules composed of an opsin protein covalently linked to a chromophore that in insects is either 11-cis-retinal or 11-cis-3-hydroxyretinal. Most insects have eyes that contain at least three visual pigments with a wavelength of peak absorbance, λmax, in the ultraviolet (UV) (300–400 nm), blue (B) (400–500 nm) and long wavelength (LW) (500–600 nm) part of the visible light spectrum, respectively, encoded by distinct UV, B, and LW opsin genes. Most of what we know about the molecular basis of vision in insects is based upon studies of holometabolous insects—insects that have four life stages consisting of an embryo, larva, pupa, and adult. In the compound eye of beetles, flies, bees, moths and butterflies, each individual ommatidium is composed of eight (i.e., flies and beetles) or nine (bees, moths and butterflies) photoreceptor cells (R1–9) that generally express only one opsin mRNA per cell, although in the eyes of some beetles, flies and butterflies, there are ommatidial subtypes in which two opsins are co-expressed in the same photoreceptor cell. Based on a phylogenetic analysis of opsin sequences from red flour beetles (Tribolium castaneum), honey bees (Apis mellifera), silkmoths (Bombyx mori), sphingid moths (Manduca sexta) and butterflies (Danaus plexippus), and comparative analysis of opsin gene-expression patterns, the patterning of the ancestral holometabolous insect eye most closely resembled the eye of the bee and the nymphalid butterfly. The R1 and R2 cells of the main retina of these insects express either UV–UV, UV–B, or B–B absorbing visual pigments while the R3–9 cells expresses an LW-absorbing visual pigment.

Following the evolution of this basic pattern for insect eyes, the visual systems of derived insects then underwent an adaptive expansion based on lineage-specific UV, B, and LW opsin gene multiplications and on alterations in the spatial expression of opsins within the eye. This pattern of opsin expansion is especially notable in the butterflies, where in every butterfly family, lineage-specific opsin gene duplications, especially of the B and LW opsins, have been detected. In the case of the fruitfly *D. melanogaster*, the majority of photoreceptor cells (R1–6) in the main retina express a 'blue-green'-sensitive (480 nm) visual pigment, *Rh1*, that is the result of an ancient gene duplication of the LW opsin but which was lost in all other studied lineages, including mosquitos. In the case of the red flour beetle, *T. castaneum*, the ancestral blue opsin was lost and an expansion of the pattern of expression of the LW opsin occurred; it included not only expression in the six outer photoreceptor cells but also in the R7 and R8 cells (equivalent to R1 and R2 in butterflies and moths), thus producing R7 cells that co-express the UV and LW opsins.

So far, opsin expression in more primitive insects remains elusive but is likely to produce additional surprises. Among them are whether or not this basic plan for photoreceptor patterning evolved early or late in insect evolution, and the extent to which sexually dimorphic eyes with respect to opsin expression pattern exists (e.g., dragonflies). Understanding the molecular sophistication and complexity of insect eyes is a challenge, which if met, has broad biological implications.

EVOLUTIONARY RELATIONSHIPS OF INSECTS

In approaching the study of the evolutionary relationships among insects and their close allies (or, for that matter, among any group of organisms), it is necessary to indicate some of the principles on which phylogenetic inferences should be based. In the past these have differed more or less radically from one group of biologists to another and it is unlikely that general agreement will be reached in the near future. However, there is growing acceptance of the view, first clearly enunciated and developed by Hennig during the 1950s, that monophyletic groups (i.e. those consisting of all the descendants of a common ancestor) are best recognized by the fact that their members share a number of specialized characteristics (synapomorphies). Shared primitive features on the other hand (symplesiomorphies) are not convincing proof of a common origin, since such characteristics may simply have been retained from more remote ancestral forms.

It follows that attempts to infer phylogeny without distinguishing between primitive (plesiomorphic) and specialized (apomorphic) features are unsatisfactory in principle. How far synapomorphic characters can be identified and how they may be distinguished from the resemblances due to convergent evolution is, however, not always easy to decide in practice, with the result that even those who accept Hennig's general principle may differ considerably from one another in the schemes of evolutionary relationship that they devise. Moreover, it seems likely that phylogenetically important features of this kind often concern obscure or relatively inaccessible structures which have not been well studied by taxonomists in the past and whose distribution among the taxa in question is therefore inadequately known. Despite these difficulties, and others which arise when one wishes to express a purely phylogenetic scheme in the form of a classification, cladistic arguments, relying on convincing sets of synapomorphic characters, seem destined to play an increasingly important part in defining evolutionary relationships. In the case of the insects this role is likely to be especially important, since the fossil record, though of great interest, is far too imperfect to allow direct recognition of ancestor–descendant connections. In what follows some attempt is made, bearing in mind the difficulties indicated, to reconstruct the relationship of the insects with other Arthropod groups and the connections of the major groups of insects (i.e. the orders) with one another.

EVOLUTION OF CIRCADIAN CLOCKS IN INSECTS

Like most organisms, insects have evolved the ability to co-ordinate their activities with the day–night cycle caused by the Earth's rotation. This has given rise to a genetically programmed timekeeping mechanism, the circadian clock, whose intracellular molecular machinery involves transcriptional feedback loops that drive persistent 24-h rhythms in mRNA and protein levels of key clock components.

Within holometabolous insects, a molecular clock mechanism has been most extensively studied in the fruitfly D. melanogaster, the housefly Musca domestica, the monarch butterfly D. plexippus and the Chinese oak silkmoth Antheraea pernyi, with more limited studies of the commercial silkworm B. mori, the honeybee A. mellifera and the beetle T. castaneum. In D. melanogaster, which has served as a model for the molecular dissection of the circadian system, the transcriptional activators CLOCK (CLK) and CYCLE (CYC) heterodimerize and bind to E-box enhancers in the promoters of the period and timeless genes, thus initiating their transcription. As the proteins are translated in the cytoplasm, PERIOD (PER) and TIMELESS form heterodimers that, after the appropriate delay in time, are translocated back into the nucleus in which PER inhibits CLK/CYC-activated transcription. Resetting of the clock by light is mediated via a blue-light photoreceptor, CRYPTOCHROME (CRY), responsible for the light-dependent degradation of TIMELESS, which ultimately resets the molecular clock's feedback loop and keeps the clock coordinated to the 24-h day.

The recent discovery of a new clade of CRYs in insects, characterized most extensively in the monarch butterfly, but also present in all non-drosophilid species examined so far (mosquitoes, butterfly/moths, beetles and bees), revealed a feature important for understanding the evolution of insects' circadian clocks. Phylogenetic analyses revealed that this new CRY family (designated insect CRY2) aligned more closely to the vertebrate CRYs, which function as the main transcriptional repressor at the core of the clock in a light-independent way. These CRY2s constitute a vertebrate-like CRY family in insects, in addition to the Drosophila-like CRY family (insect CRY1). Based on phylogenetic analyses, these two families of insect CRYs appear to be the result of at least two rounds of gene duplication at the base of the metazoan radiation. Functional assays in cell culture have shown that insect CRY2s are light insensitive and are potent transcriptional repressors of the clockwork, as are their orthologs in vertebrates, whereas insect CRY1s are light sensitive with no transcriptional repressive activity.

Therefore, within insects, it appears that a single clockwork mechanism evolved that has undergone specialized changes in various lineages, through the processes of gene duplication and loss to produce at least three types of clock mechanisms: (1) One form is the ancestral clock (apparent in the monarch butterfly, silkmoths and mosquitoes) in which both insect CRY1 and CRY2 exist and function differentially within the clockwork; insect CRY1s function in light-entrainment of the clock, while insect CRY2s function as the main transcriptional repressor of the clockwork; (2) Drosophila has a derived clock in which the vertebrate-like CRY2 has been lost and Drosophila-like CRY1 functions primarily as a circadian photoreceptor in clock neurons in the brains of flies; and (3) Beetles and bees also have derived clocks in which the light-sensitive CRY1 has been lost and only the insect CRY2 exists and functions within the clockwork.

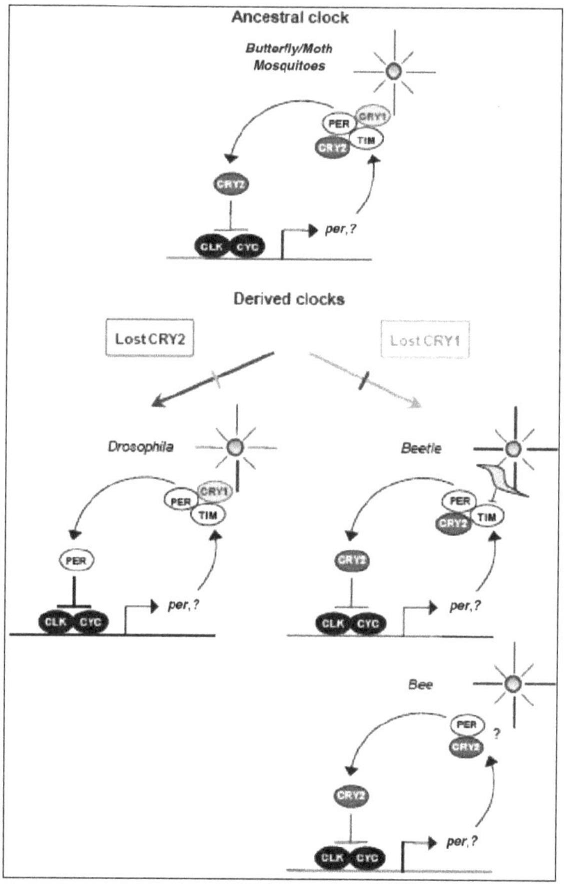

Evolution of insect clockwork mechanisms. The ancestral clock (butterfly/moths and mosquitoes) possesses the two types of cryptochromes (insect CRY1 and insect CRY2). Gene losses gave rise to two types of derived clocks, one possessing only CRY1 (Drosophila) and one possessing only CRY2 (beetle and bee). CLK, clock; CYC, cycle; PER, period; TIM, timeless. The sun represents the light input to the clock. In the derived clock found in the beetle, the flag represents a putative photoreceptor involved in TIM degradation for clock entrainment, and in the bee clock, the exclamation point represents a novel light input pathway to the clock.

Drosophila, as a genetically tractable model, has been invaluable for understanding the molecular basis of the clockwork mechanism in animals. However, studies in other insects, like the monarch butterfly, have provided an "ancestral model" for comparison of clockwork mechanisms among insects, and between insects and mammals. Importantly, the recent development of comparative analysis of the clockwork mechanisms within holometabolous insects would not have been possible without the availability of their genomes. With the recent explosion in genome-sequencing methods, extension of molecular studies to the circadian clocks of hemimetabolous insects (e.g., cockroaches, locust, and crickets), in which much physiological and anatomical analyses have been performed, should be forthcoming.

EVOLUTION OF EUSOCIAL INSECTS

The repeated origins of eusocial organisms can be considcred one of the major evolutionary transitions in the history of life. Most eusocial animals are insects, and several lineages including termites, ants, wasps, and bees have evolved diverse forms of social organization and often display considerable ecological dominance. Eusociality traditionally has been defined to include three key criteria: generation overlap, cooperative brood care, and reproductive division of labor. These societies can also be viewed as displaying high reproductive skew where relatively few individuals monopolize reproductive output, leading some to consider as eusocial gall-inhabiting aphids and thrips, and one species of beetle. The major lineages of eusocial insects, however, are termites and several groups within the order Hymenoptera (ants, bees, and wasps) that provide the focus of the present survey.

Ants (Formicidae) comprise the largest eusocial insect lineage in terms of described species (<12,000) and often are ecologically dominant members of their communities. All ants are eusocial, and the origin of eusociality is clearly quite old in this group. The most recent molecular divergence-dating analysis estimates an age of 115–135 Ma for crown group ants, and thus eusociality must have originated sometime before that. Phylogenetic work shows that some taxa previously considered to reflect early ant eusociality, such as the bulldog and dinosaur ants of Australia, instead are nested well within the phylogeny of ants. Several subterranean lineages with cryptobiotic characteristics resolve at or near the base of the ant tree in several studies. It remains possible, however, that the closest stem-group fossils, the Sphecomyrminae, which were large-eyed and probably lived in exposed environments, better represent the groundplan for early ants. Ants throughout their history have used eusociality as a platform on which many behavioral specializations have evolved. A particularly astonishing example occurs in the fungus-growing ants, the most important group of agriculturalists other than humans. Recent molecular phylogenetic and divergence dating analyses provide a historical timescale for this symbiotic system over the past 50 Ma and reveals that leaf-cutting ants, the dominant herbivores of the Neotropics, evolved remarkably recently, only 8–12 Ma.

Within wasps of the family Vespidae, eusociality occurs in the Polistinae (paper wasps), Vespinae (yellowjackets), and Stenogastrinae (hover wasps). Morphological and behavioral characters argue that these three groups form an exclusive clade, indicating a single origin of eusociality. A recent molecular phylogeny, however, suggests that Stenogastrinae is only distantly related to the other two eusocial groups, implying two eusocial origins. Several trace fossils of wasp nests are preserved throughout the Cenozoic, the oldest of which is a presumed nest of Polistinae or Vespinae at least 63 Ma, establishing a minimum age for a eusocial origin in this group.

Eusociality also occurs in at least one species of *Microstigmus* wasp in the family Sphecidae and may extend to related species and genera, although these taxa remain poorly studied.

The remaining instances of insect eusociality occur within the lineage Anthophila (bees). The vast majority of the ~18,000 described species of bees are solitary or cleptoparasitic, with only approximately 6% displaying eusociality. Molecular phylogenetic analyses clearly indicate that a solitary lifestyle is the ancestral state for bees, and that eusociality evolved multiple times within bees. Eusociality occurs in three groups of bees, corbiculates, allodapines, and halictids.

The corbiculate bees are divided into four tribes: Apini (honey bees, including *A. mellifera*), Meliponini (stingless bees), Bombini (bumblebees), and Euglossini (orchid bees). All tribes except Euglossini are eusocial, with Apini and Meliponini displaying especially elaborate eusocial colonies. In a situation similar to that in vespid wasps, morphology and molecules conflict over the phylogenetic resolution of the corbiculates. Morphological studies largely support a phylogenetic arrangement of {Euglossini, [Bombini, (Apini, Meliponini), implying a single origin of the advanced eusociality seen in Apini and Meliponini. However, ever-increasing molecular data often favor the alternative view that Bombini and Meliponini form a sister group to the exclusion of Apini, challenging the notion of a single origin of advanced eusociality. The oldest known crown-group bee fossil at ~65 Ma (*Cretotrigona prisca*) falls within Meliponini, placing a lower bound on the origin of eusociality in this group.

Eusociality also occurs in the allodapine bees, one of four tribes within the sub-family Xylocopinae. Allodapines show a broad range of eusocial behavior and recent work has demonstrated that no allodapine group is strictly solitary. Molecular divergence-dating studies suggest a single origin of eusociality in this group dating back 40–80 Ma.

Halictidae (sweat bees) exhibit particularly flexible forms of eusociality over both ecological and evolutionary timescales. Eusociality occurs in three halictid taxa: Augochlorini, *Halictus*, and *Lasioglossum*. These groups have experienced several evolutionary reversals back to a solitary state, as many as 4–6 times both in *Halictus* and *Lasioglossum*. Molecular divergence dating infers recent origins for halictid eusociality, ~20–25 Ma in each of the three groups. This time period coincides with a global warming trend during the late Oligocene warming and mid-Miocene climatic optimum. A potential correlation between eusocial evolution and climatic warming is strengthened by observations that climatic factors influence the manifestation of eusociality in some modern species that have solitary forms in colder areas and eusocial forms in warmer areas; this switch may be due to an increased growing season necessary for producing multiple seasonal broods. These results combine to suggest that additional study of the taxonomy, phylogeny, and social evolution of halictid bees may provide an especially valuable window into the origins and early maintenance of insect eusociality.

Conservation of Insects

CHAPTER 6

The process of caring and preserving insects through collection, preparation, storage, examination, documentation, research and treatment is referred to as insect conservation. Its psychology and practices are also studied within it. This chapter sheds light on the subject of conservation of insects for an in-depth understanding of it.

Insects are the most speciose organisms on earth, making up 70% of all organisms. They dominate all but the coldest and saltiest environments. They inhabit deserts to tropical forests, and swampy pools to pounding waterfalls. They are the majority that few of us see, hidden in plain view. All terrestrial and freshwater plants, even mosses and liverworts, have associations with insects. Most plants have flowers, and their reproduction depends on insect visitors to pollinate them, and so reproduce. Virtually, all frogs and lizards need insects to sustain them. Well over half of all fish, birds, and small mammals require insect food. In turn, a third of insects eat other insects. In short, insects are the fundamental woof and weft of all land-associated ecosystems. Furthermore, we cannot live without them, as a third of our food, and especially the most nutritious components of our food, such as fruit and nuts, depends largely or totally on insect pollination.

Insect success has come about largely through the insect's body plan, with its three tagma (head, thorax, and abdomen), its immensely versatile skeletal structure, and highly varying physiology. The head is packed with a huge array of sensory apparatus. The thorax has highly effective legs for many environments, and, most notably, wings for dispersing and rapidly finding resources and mates. In turn, the abdomen, houses diverse digestive tracts, as well as reproductive apparatus that in some species produces millions of eggs. Indeed, reproductive potential can be extraordinarily high. Richard Harington calculated that when a gravid aphid is left to reproduce with no mortality, after 1 year, the earth would be covered 14.7 km deep in aphids.

Insects have a wide range of mouthparts molded from the robust chitin of the skeleton, so that they can chew, rasp, suck, and burrow through all sorts of organic tissue, living and dead, plant, animal, and fungal. They are also able produce an immense array of chemicals for attack, defense, camouflage, mate attraction, and digestion. Furthermore, their sensitivity to certain chemicals can be extraordinary, with some moths being able to detect a mate many kilometers away, and others, such as certain parasitoids, able to detect prey deep in plant tissue.

Insects are not just items, but also interactors. They are among the most ecologically connected of all organisms. A simple biotope with just 1000 species, leads to half a million potential interactions. This means that the over 1 million described insect species and the likelihood that there are about 5 million species in all, suggests that insects interact with virtually every component in the terrestrial and freshwater realms.

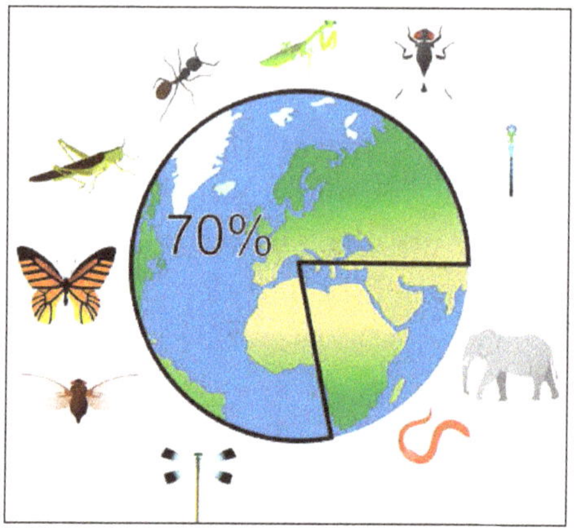

70% of the species on Earth are insects, despite the land covering only a third of the planet. Insects in terrestrial and freshwater systems are the most highly ecologically connected of all organisms.

Insects in Today's World

The world is in trouble, with the "World Scientists' Warning to Humanity: A Second Notice" having been issued. We must act now, and decisively, on how we manage the planet. The Anthropocene ("age of humans") is well-established as the new geological era, and the sixth mass extinction is upon us. Insects are central to how we react to this crisis, and how we should respond, as avoiding general ecocide in the twenty-first century rests on involving insects in the new world view.

What is of great concern is that insects appeared to have pulled through the last great extinction at the end of the Cretaceous, 66 million years ago, largely intact. They also survived the various glacial maxima and minima by moving around to re-establish in thermal optima, and to some extent, independent of plants. Such large-scale movement is not so feasible today. The human-induced patchwork of anthropogenic, novel ecosystems has created a myriad of barriers to free movement. With global climate change and landscape fragmentation being a "deadly anthropogenic cocktail", the future for insect diversity depends on three options: (1) adapt on site, (2) move across the human-instigated barriers, or (3) die out. As (3) is not an ethical or survival option, either for insects or us, we must find ways that enable insects to survive through the twenty-first century and beyond.

Despite the importance of insects, it is only relatively recently that they have been mainstreamed into biodiversity conservation. This is being done at various operational levels from the species level through to the landscape level of conservation, with major decisions being made at the scale of nation states through National Biodiversity and Strategy Action Plans. Furthermore, as there are now global insect conservation initiatives, it highlights the adage "think global, act local." These operational levels, species, landscape, national, and global, are not mutually exclusive, but complementary.

Insect conservation in the twenty-first century can be seen against six inter-related themes: (1) philosophy (establishing the ethical foundation), (2) research (the finding out), (3) policy (the framework for action), (4) psychology (understanding how humans engage in insect conservation action), (5) practice (implementation of action), and (6) validation (establishing how well we are doing at conserving insects).

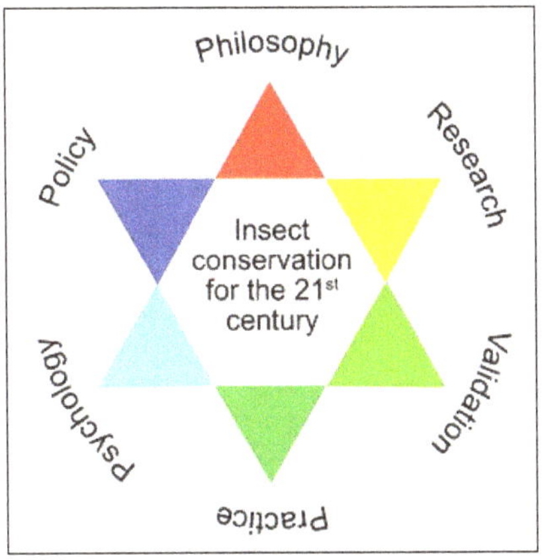

One way to view insect conservation into the twenty-first century is to focus on six, inter-related themes: philosophy, research, policy, psychology, practice, and validation.

Philosophy for Insect Conservation

The starting point for insect conservation is to question why we should do it. Arguably, as extinction is the norm, with 99% of all organisms on earth having gone extinct from natural causes, perhaps we should just let events take their course, as a part of an evolving planet? There are two aspects here that we must consider to counter this view. First, there is the intrinsic value of insects, and that they must be conserved for their own sake, especially as they celebrate the immense complexity of life. The argument for intrinsic value is that we are sentient, and hopefully, as we have given ourselves the epithet *sapiens*, a wise and caring species. Quite simply, we share this lonely planet with an amazing variety of life and a stunning selection of insect forms.

With humanity having received its "Second Warning" and global ecosystems in major decline, we cannot carry on as we have been up to now. We require a radical change in thought and action. We need a good philosophical base for steering practice. Many people consider that conservation is instrumental and must benefit humans. This approach stems partly from the logic that by taking this approach, those who hold power will listen. In short, it is considered by many as the only hard currency of insect conservation.

The binary approach of intrinsic versus instrumental value of itself has shortcomings, as it focuses on entities, such as insect species or landscapes, rather than how we relate to nature. There is now a move away from this binary approach to one that focuses on personal and collective wellbeing, based on how we value and relate to nature to achieve this wellbeing. This focus on relational value is built into our need for nature, and that we have a shared destiny, with biodiversity as a whole. This also means relating to insects as most of them are fundamental to our health and happiness, because without them we would have an impoverished and dangerous world as resources decline. Quite simply, we need to look after insects, and they will look after us. We can no longer ignore this fact if there is any future for our grandchildren.

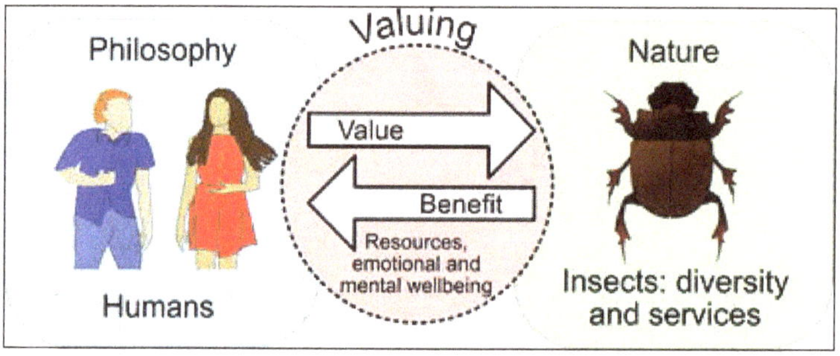

Justifying insect conservation, and then doing it requires a philosophical view based on valuing nature. When value is placed on the human relationship with nature, benefit accrues in terms of physical resources such as food, as well in terms of well-being, both emotional and mental.

Research needed for Twenty-first Century Insect Conservation

Operational Levels of Insect Conservation Research

Research is concerned with discovery of new information. For insect conservation, this research is about finding new and effective ways for maintaining insect diversity, insect species, and insect populations. As insects are embedded in the ecological fabric around them, and we need to understand it if we are to provide realistic insect conservation solutions, we research the optimal environmental conditions that enable insect survival. These environmental conditions may be abiotic, such as temperature regimes, fire frequencies and intensity, rainfall patterns and intensity, insolation, elevation, rockiness, water, pH, dissolved oxygen, as well as contaminants, pollutants, pesticides, and many others. Environmental variables are also biotic, including vegetation structure and composition, pollen and seed availability, fungal presence vs. absence, host availability (vertebrate or invertebrate, including other insects), mutualist presence, dung availability, mimic models, and so on.

As many insects express developmental polymorphism, where for example, the larva is morphologically and functionally very different from the adult, they may require a host of abiotic and biotic conditions and resources for optimal survival within their habitat. The larva of a butterfly needs a particular host plant(s), as well as enemy-free and disease-free space, while the adult needs certain nectar sources, mate-meeting sites such as hilltops, oviposition sites, sunny conditions for flight, besides enemy- and disease-free space.

Although an insect's habitat is embedded in an ecosystem, some require more than one ecosystem to sustain them. With anthropogenic modification of the landscape, not only are conditions changed within their habitat, but also around it. The landscape matrix around a habitat may not only lack critical conditions and resources, so prohibiting generational survival there, but it also has an effect, often adversely, on the habitat itself. In terms of research, as well as investigating the habitat per se, we also must establish the landscape context and contrast. How we make this matrix more hospitable for insects is a major research thrust in this the twenty-first century. This is where we need to reconcile the needs of insects and those of humans. Progress is being made, but now we must hasten that process in the coming years.

Insects have to move and mate, and so maintain genetic diversity within a species, especially to maintain adaptability to changing conditions induced by humans. Functional connectivity across the landscape that facilitates movement, and so genetic exchange, has become a major challenge for the twenty-first century. Already much progress has been made. This has been done by making the human production landscape and urban environment more insect friendly. This has been done by putting in place stepping stone habitats, while perhaps not being optimal for long-term survival nevertheless provide stop over stations for insects as they move across the landscape, and by instigating conservation corridors, which as well as being for movement, are also source habitats that provide optimal conditions for all the life stages and the production of viable offspring.

Species-level Insect Conservation

From a conservation perspective, we can view "a species" as populations made up of a group of individuals. We do not actually conserve "a species" but individuals of the species. Groups of individuals in an overall population do, or do not, exchange occasional alleles. Those that do are metapopulations, and those that do not are subpopulations. Importantly, it is the range of genetic variation in populations and how it is shared among individuals that determines the adaptability of a population to environmental change, whether for the better or the worse, and whether in response to local (e.g., landscape fragmentation, pollution) or global (e.g., climate change) impacts, or both.

The viability of metapopulations depends on the flow of genes that provide high value for adapting to prevailing conditions. These conditions are currently changing rapidly, and are often adversely synergistic, with, for example, fragmentation and insecticidal impact together providing an even greater challenge over the impact of just one. Without gene flow, metapopulation dynamics can be disrupted, leading to an adapt-or-die situation. There might not be enough genetic variation within the isolated group of individuals for survival in the long term, because of changing or stochastic environmental events and genetic impoverishment. When populations are isolated by the effect of landscape fragmentation, there may some selection pressure to the new conditions, leading to human-induced evolution, known as anthropovicariance. Philosophically, this leads to an interesting debate. Here, we are not conserving an existing natural phenomenon so much as creating a new one: To what extent is this "new insect" of conservation concern?

Subpopulations present a different situation in that they already show some differentiation caused by natural drivers. These different subpopulations are known as evolutionarily significant units (ESUs), each unit of which deserves conservation in its own right. They are in effect evolution in action, and represent new species in the formation. However, some ESUs are threatened and others not. The English Large copper butterfly *Lycaena dispar dispar* is extinct, the Dutch ESU *L. dispar batavus* is highly threatened, and the Estonian *L. dispar rutilis* is common and expanding its geographical range. The "species" has been re-introduced and has established in England from Europe, but this is not the original ESU. These are genetic and ethical issues that will confront us ever more this century.

The twenty-first century is likely to see much more focus on the genetics of species, bearing in mind this will always be about a few species attracting special attention. There will be several approaches, and these are already developing. Genetic work has shown that some species are very ancient, with the yellow presba dragonfly *Syncordulia gracilis* having a pedigree going back almost 60 million years. Such species must receive conservation action if we are to show some empathy for ancient insects. Even resurrecting extinct species (revenant species) is feasible from well-preserved specimens in museums. However, this is a lot of hard work for a privileged few in comparison with saving species by good and protective management of natural ecosystems in the first place.

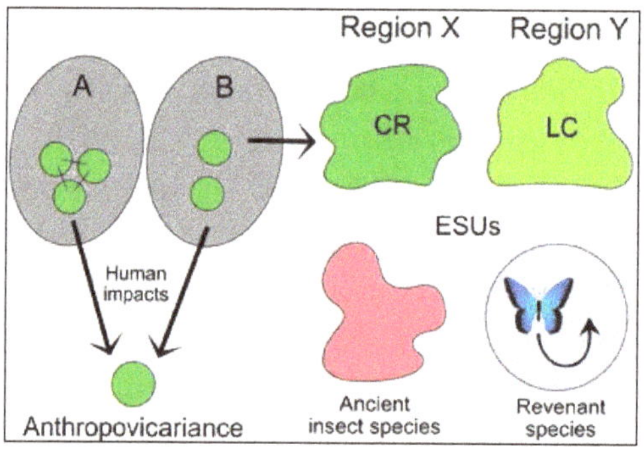

There are great opportunities for insect species-level conservation, as long as we have a good understanding of what "a species" stands for, based on genetic, behavioral, and ecological knowledge. Here (upper left) are two species (A and B), one (A) with populations that experience genetic exchange (metapopulations), and one (B) that does not (having two subpopulations). Either of these might undergo genetic change in response to human impacts, a phenomenon known as anthropovicariance (bottom left). Two subpopulations may be genetically, and usually morphologically, different, making up two evolutionarily significant units (ESUs) in two geographically distinct regions (X and Y) (upper right). The one on the left is red listed as critically endangered (CR), and of great conservation concern, while the one on the right is red listed as least concern (LC), which means that it is not of immediate concern, but it could always be threatened in the future. Ancient species, with a long phylogenetic pedigree, are often of high conservation significance as they are usually genetically highly irreplaceable (bottom middle). A revenant species is one that is extinct as a species and has been brought back to life from good quality genetic material extracted from museum specimens (bottom right).

Landscape-level Insect Conservation

Species conservation is arguably a luxury overlay on insect diversity conservation using landscapes, at least from a global perspective. Insect species conservation is morally right, but exclusive, and time is short for conserving as many insect species as soon as possible. However, there is a good reason to do good insect conservation in parts of the world that can afford that luxury, i.e., in those countries with high GDPs, and where there is great interest and involvement by the public. It leads to exploratory techniques and methods that we will, in the future, need globally. Meanwhile, for those countries with lower GDPs, as well as the more economically developed nations, conserving good quality landscapes with high habitat heterogeneity will not only conserve many insect species and their interactions all at the same time, but also will conserve many insect services to which the public and policy makers can both literally buy into.

Conceptually, we as humans relate to landscapes very well, as they fit comfortably into our frame of reference about nature. We see this already with the evolution of new perspectives. While there has been much progress in the past with species conservation, there is now a shift towards viewing nature as a vast array of benefits that it provides, which includes being in nature for our wellbeing.

Natural England's Conservation Strategy for the twenty-first century articulates this well, and uses three guiding principles: (1) creating resilient landscapes (and seas), (2) putting people at the heart of the environment, and (3) growing natural capital (i.e., giving populations the chance to survive and increase). The earlier strategy of ring-fencing and protecting individual species and habitats has not been successful, having led to local species loss. The focus now is at the larger spatial and conceptual scales, with the development of resilient landscapes and ecosystems. This has led to research on the drivers of species loss and deterioration of ecosystems vis-à-vis what maintains species, their interactions, and ecosystem function. This means understanding what are opportunities and realistic strategies can be brought into play. These include engaging wildlife-friendly farming, gracing the urban environment with biodiversity-friendly green spaces, and improving functional connectivity across the landscape.

All these biological perspectives must include people and their wellbeing if it is to succeed. The focus moves away from risk toward a new approach that involves enhancing and investing in the environment, leading to long-term stewardship of environmental assets. Interestingly, this new approach does not exclude a species focus, but rather integrates it into a vision of long-term resilience across landscapes. Crucially here, it includes insects of all types, whether threatened or not, which contribute to a more informed landscape approach. It is essentially a positive feedback loop, with all species, including humans, benefitting. In turn, it is the sensitive species on those landscapes that can tell us how well we are doing. These principles now require more research to tailor them to local circumstances.

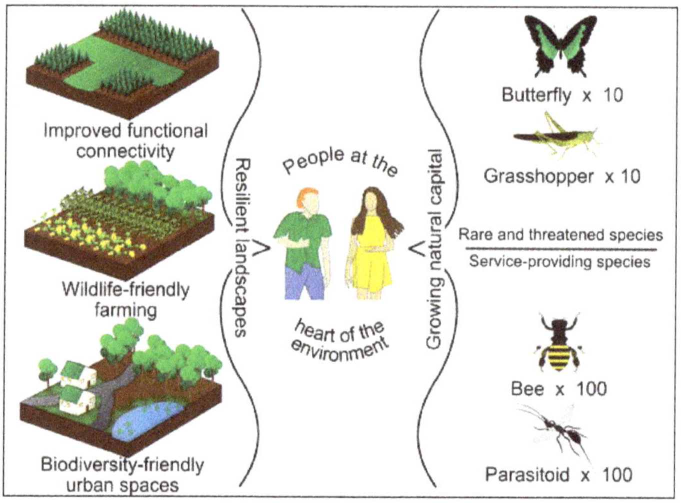

Landscape-level insect conservation based on three guiding principles: (1) creating resilient landscapes, (2) putting people at the heart of the environment, and (3) growing natural capital, as perceived by Natural England. Resilient landscapes include, for example, improved functional connectivity (here, grassland conservation corridors between exotic pine plantation tree blocks, wildlife-friendly farming such as using organic methods and leaving an increased proportion of natural habitat, as well as biodiversity-friendly spaces in the urban environment). Growing natural capital is just not only about improving indigenous species richness, but also improving the abundance of the focal species. In the case of rare and threatened species, an increase of ten times would be a great improvement, while for service-providing species, like bees for pollination and parasitoids for pest control, a 100-fold increase would be truly excellent.

Policy for Insect Conservation

There are two ways to consider insect conservation as regards policy. Insects are a component of biodiversity, and secondly, they provide essential ecosystem services. These perspectives can be considered at various spatial scales, from global down to national (bearing in mind that nation states are an important conservation action unit), and then local.

From a biodiversity perspective, insect conservation is integral biodiversity conservation, as insects function at various trophic levels, and so interact with many other organisms, plant, fungal, animal, protoctist, as well as among themselves. Biodiversity conservation is globally framed in terms of the Convention on Biological Diversity (CBD), which is agreed to by most nation states. Importantly, the Aichi Biodiversity Targets, under the umbrella of the CBD, provide some specific goals, and with the exception of those specifically relating to the marine environment, apply to insect conservation into the future. There are, however, a few marine insects, notably sea striders (Hemiptera, Gerridae: *Halobates* spp).

Insects globally are conserved both at the species level and as insect diversity. At the species level, it is the International Union for the Conservation of Nature (IUCN)/Species Survival Commission (SSC) that is the major global proponent through production of the Red List of Threatened Species. This globally recognized list provides just not only an inventory of the world's species and their threat status, but also gives suggestions for conservation action. It does this through the activities of a network of specialists on various taxonomic groups.

The red list however, is not a priority list for action, and nor is it a political document stipulating what governments or agencies should do to conserve insects. Quite simply, but importantly, it is an assessment of the conservation status of the insects that have so far been assessed. Nevertheless, it is highly influential for conserving insects globally. One reason for its impact is that once a species is assessed on the red list, and especially when a species in listed as threatened (as opposed to not, i.e., that it is classified as Least Concern), a species tends to become highly iconic. This is a great boost for listed insects, as they then receive the same treatment as a wombat, whale, or weasel. In short, their profile is raised considerably, and so automatically find their way into policy documents on biodiversity conservation.

The greatest challenge, among many, for insect species Red Listing is that the group is so speciose, with today only about 7700 species having been evaluated for the Red List, which is 1% of described species, 1,060,704 in all, and probably less than 0.2% of the millions that exist. The reason for these low percentages is that considerable field work is required to assess the threat status of an insect species, and there are relatively few insect specialists to do the job. The situation is aggravated by many species going extinct without even having received scientific names, a phenomenon known as Centinelan extinction (named after Centinela Ridge in Ecuador where botanists found many plants had gone extinct from deforestation before they could name them).

Of those insects that have been assessed, it is possible to get a sense of the types of threats facing them. When lumped with terrestrial invertebrates in general, the main threats in decreasing order are: (1) habitat loss due to logging, (2) habitat loss due to agriculture, (3) infrastructure development such as urbanization, (4) habitat loss and fragmentation due to transportation/service corridors, (5) invasive alien species, (6) change in fire regime, (7) pollution, (8) climate change/severe weather, and (9) mining. For freshwater invertebrates, pollution and dams/poor management are the major threats, and greater than all the others together. Knowing what these threats are not only helps us plan for the future, but also bearing in mind that for any one insect species, there may be more than one threat, as threats are often adversely synergistic.

International trade in insect species is regulated. This is done through the Convention on International Trade in Endangered Species (CITES). To date, however, few insect species are listed, and they are mostly large, highly collectable and charismatic species, like birdwing butterflies. Objectively, there should be many more insect species that are CITES listed, but this is not the case because the limited financial resources for doing so are prioritized for restricting trade on charismatic vertebrates and rare plants, which are facing a desperate plight in its own right.

Moving down to the national scale, insect species conservation varies greatly from one country to another, with those countries with a high Gross National Product usually devoting more attention and resources to insect species conservation than the financially more constrained countries. Some countries have strong regulation protecting insects (e.g., all dragonflies in Germany). There are also consortia, and especially notable is the European Union Habitat Directive, where there is co-operation on protecting insects across the continent.

As regards, ecosystem services as defined by the Millennium Ecosystem Assessment, insects are highly significant. In terms of provisioning services, they are important as biological control agents of pests, for monitoring ecosystems, as well as for providing new medicines, and acting as tourist attractions in the wild, and in commercial butterfly houses in urban areas. Regulating services provided by insects include nutrient cycling, pollination, seed dispersal, stopping, or slowing invasions by other insects, and contribution to atmospheric gases. They also provide supporting services through breaking down living and dead plant material, as well as turning over soil. In turn, cultural services are many, and include representativeness of the variety of life, connecting with the natural world (especially through children), and use in genetic research.

All these services are often strongly recognized by policy makers, and regularly form the basis of justification for conserving insects, i.e., for what they do for us, which in the USA is estimated to be $57 billion/year. But, we must be careful not to confuse the conservation of insects for the services that they provide with the conservation of insect species, which may not be the same set.

With recognition of the value and importance of insects with each passing year, insects are being increasingly enshrined in policy, at least in countries that have great respect for nature and recognize that our survival depends on valuing and conserving them. Recognition of the value and importance of insects will become more important with the passing of time this century.

Insects have been eaten by humans (entomophagy) from the earliest of times, with today an estimated 2 million people eating insects as part of their traditional diet. This is because insects are highly nutritious, being high in protein, essential fatty acids, and in important minerals such as calcium, copper, iron, selenium, and zinc. Commonly eaten insects are beetles, moth larvae, bees, wasps and ants, cicadas and other bugs, termites, dragonflies, flies, as well as some other insects. Entomophagy will become increasingly important in the twenty-first century, both for direct human consumption and for livestock. Entomophagy is also now being used as a way of tackling malnutrition in children.

This increased reliance on insects is partly driven by red meat production being three times more resource hungry than insect production. While insects traditionally have been harvested from the wild, there is now a move to rear them on a large scale. This intense farming of insects has challenges, but new rearing and processing techniques and methods will inevitably come about simply because natural habitats are decreasing, and the size and demands of the human population are increasing. Improved insect farming will go hand in hand with new developments in the visual appearance of the food, and its design, so as to make insect food increasingly acceptable to a more discerning human population.

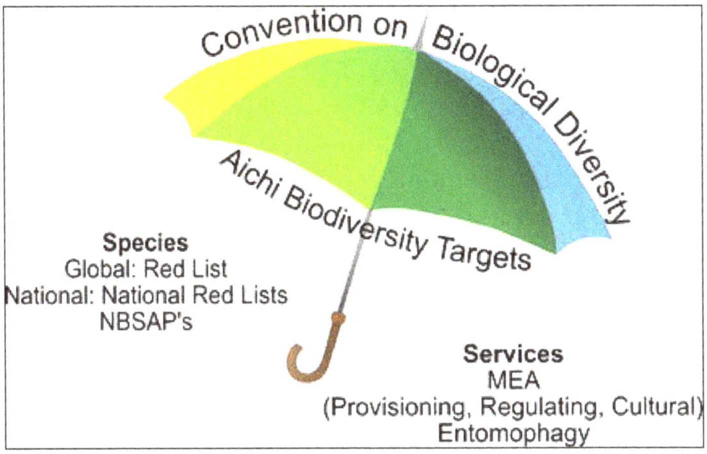

Insect conservation policy has various perspectives. Overarching is the Convention on Biological Diversity to which most countries in the world are signatories. The Aichi Biodiversity Targets, which are specific targets, mostly have great application to insect conservation. In terms of conservation of species per se, the International Union for the Conservation of Nature's global red list gives important coverage to insects. At the national level, local red lists (RLs) are important too. National Biodiversity Strategy Action Plans (NBSAPs) are also significant for local insect conservation, especially in biodiversity hotspots. NBSAPs cover a wide range of biodiversity issues and targets besides insect species conservation. As regards services supplied by insects, these may be described under the Millennium Ecosystem Assessment (MEA). It provides a framework to include provisioning, regulating and cultural services. An important service currently being supplied by insects is entomophagy, the human consumption of insects, which is becoming increasingly mainstream globally.

INSECT CONSERVATION PSYCHOLOGY

Understanding human behavior, promoting wellbeing, and increasing human care for nature is conservation psychology, which recognizes our dependence on nature and an understanding of why and how it is so essential for our wellbeing. Given, the huge role that insects play in providing services of all sorts, insect conservation psychology is going to play an important, if not vital, part in our survival into the twenty-first century.

As we advance technologically, we seem oblivious of the collateral damage we are doing to so many other species and their interactions, despite so many warnings that we are vitally dependent on them. In short, we are doing untold damage to the very systems on which we depend, with insects being at the heart of them, at least on land and in freshwater. We simply take nature, and insect services, for granted, and this cannot continue into the twenty-first century without dire consequences. Just as we do not see water vapor coming off plants, or the oxygen being breathed out by them, we just do not see insect diversity and our dependency on it: out of sight, out of mind.

Many of us have great interest and even respect for insects as children. But there are only certain "species," like "the" butterfly, bee, grasshopper, ladybird, and dragonfly that we hold dear. These are the iconic insects that are fascinating to us as children (and adult entomologists!), and are also benign. Only as we grow up, do we begin to realize that "the fly" is dirty, and "the mosquito" is a nuisance, if not dangerous. So, all insects get lumped together in our adult perceptions, i.e., not worthy of our consideration and conservation. This realization is now changing for the better, with the global realization that we have a pollinator crisis around the world, and "the bees are dying". This has aroused considerable awareness of our dependency on insects.

It has also dawned on us that we need to be far more aware that widespread use of insecticides is not only detrimental to our health but also undermining our very food base, despite actually having been articulated in the 1960s. The pollination crisis has also led to much action to find solutions to it, and for once, right now without delay. Like climate change, the pollinator crisis is no longer something to think about in the late twenty-first century—it has already arrived with full force.

While our dislike of many insects is intrinsic (biophobia/entomophobia), love and appreciation of insects (biophilia/entomophilia) is learned. This learning can only come about from positive experiences, for example, by seeing insects in their natural environment, whether in the home garden, botanical garden, or protected area. Indeed, protected areas receive 8 billion people visits per year, with a profit of $250 billion/year, making protected areas, along with other natural areas, good for human wellbeing, while also earning their keep and supporting a vast array of insects.

Linking insect conservation psychology to research and policy can come about through citizen science, the involvement of informed and enthusiastic sectors of the public for recording the distribution of species and engaging in insect conservation. This involvement of the public has greatly improved our knowledge of many insect groups. The improvement in knowledge, and hence insight, comes about simply because there are more enthusiastic eyes and hands in the field increasing the amount of information on species, and this can be very high in the case of threatened, red listed species, as well as many others that are not threatened, at least for now. This information then also becomes valuable for lobbying policy makers. Enthusiastic citizen scientists are also making direct contact with policy makers, and therefore increasing the voice of change-for-better in insect conservation. This approach is going to become increasingly important in the twenty-first century.

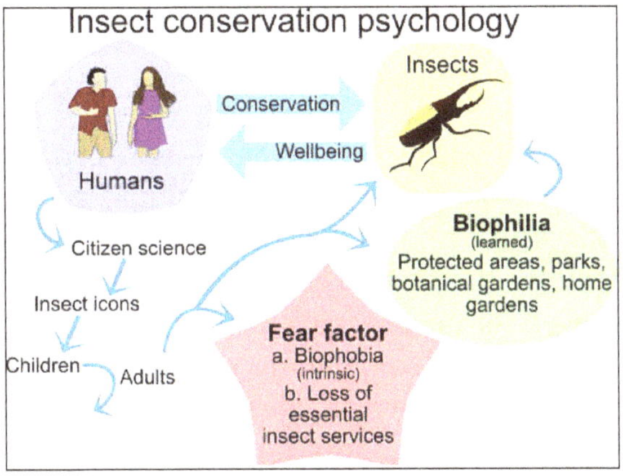

Insect conservation psychology is the relationship between humans and insects. The conservation of insects is essential for our wellbeing. The involvement of the wider body of the public in recording, monitoring, and engaging in conservation is citizen science. This activity promotes wellbeing in people while helping ensure a future for insects. Insect icons like "the bee, butterfly, and ladybird" are highly significant for children, but often we as adults forget them in our preoccupied lives. This is partly because an awareness develops that "the fly and the mosquito" are not good for us. We are also aware that the wasp stings, and so will the bee if we are unkind to it. This is biophobia (a), which is intrinsic to us. Yet when we overcome this fear factor, we culturally develop biophilia, especially when we see the beauty of nature, including insects, in our parks and gardens. Besides biophobia, the fear factor today has a second component, a concern that there is a loss of essential services (b), particularly pollination. Our reaction to this service loss, alongside biophilia for certain species, is feeding back into insect conservation, which, in turn, improves our sense of wellbeing.

INSECT CONSERVATION IN PRACTICE

Insect conservation is a practical activity based on good research and a sound knowledge base. The foundation for practice is a well-understood philosophical underpinning, valuable research findings, and then a strategic response to well-informed policy, alongside applied insect conservation psychology. Yet, the practice of insect conservation is rarely actually carried out by philosophers, researchers, policy makers, or psychologists. Insect conservation action is usually carried out by conservation organizations, concerned farmers, and progressive private companies, and other responsible land stewards and owners, all of whom are highly cognizant of philosophy, research, policy, and psychology. It is also carried out by citizen scientists, many of whom may be associated with self-grown non-governmental organizations like the Xerces Society, Butterfly Conservation, and Buglife, who receive donations from private funders, as does IUCN at the global scale. Citizen scientists are usually active in their home area, where they carry out insect conservation activities in addition to monitoring, especially on nominated insect species, and often those of special concern. This is another reflection of the "think global, act local" principle.

There is a baseline for insect conservation that is non-negotiable, that is, the recognition of formally proclaimed protected areas composed of natural habitat. The significance of this is that these areas are often an island in a mosaic of novel landscapes. They are frequently the last bastion for many rare, specialist and highly localized endemics, as well as for specialized interactions. As protected areas are usually isolated, they may require some management, usually to mimic the historic condition. Introduction of fire regimes, for example, may be required to simulate the situation before modern human fragmentation of the landscape. Management of fire may be to reduce the fuel load, or so reduce risk of an intense and potentially highly damaging fire, or it may be to stop vegetation succession to an unnaturally woody environment.

There are approaches that also intergrade protected areas with the surrounding novel mosaic. Among these are biosphere reserves, of which there 669 in 120 countries, and ratified by the United Nations Educational, Scientific, and Cultural Organization. Biosphere reserves consist of a core and two surrounding zones. The core of the biosphere reserve is fully protected and managed according to historic conditions. Surrounding that is a buffer zone with only low-level human impact on the biota and ecosystem function. Outside the buffer zone is the transition zone, which supports sustainable agriculture or forestry based on agro-ecological principles or organic agriculture, as well as having little infrastructure development.

The transition zone is characterized by low insecticide use, maximal use of biological control, and cultural control measures of pests, low compaction practices (i.e., avoiding heavy machinery), planting of indigenous flora, and the setting aside of remnant patches as reserves per se or as stepping stones to protected areas. Despite the great importance and high level of instigation of biosphere reserves, there is still little research on their effectiveness for conserving insects and their interactions. Research and validation of these reserves is going to be increasingly important in the twenty-first century, as they may well prove to be a major initiative for harmonizing human activity for optimal production and effective insect conservation.

(Left) An organic vineyard with no insecticide input, mulching of the inter-rows, and the planting of the inter-rows to biodiversity-friendly vegetation. These vineyards are particularly rich in soil fauna as well as in above-ground insect diversity. (Right) Seen here is a large-scale ecological network of conservation corridors of remnant, natural, high value grassland in and among plantation forestry using alien pine trees. These grassland corridors not only conserve biodiversity but also maintain hydrological processes in a natural state.

Insects Interactions

7

CHAPTER

The relationship between insects and humans fits into all three categories of a biotic relationship i.e. parasitism, commensalism and mutualism. Negative impacts of insects include transmitting diseases, damage to structures, and positive effects consists of pollination, creating valuable goods like silk, lacquer, dyes, etc. All the aspects of this relationship between insects and humans have been carefully analyzed in this chapter.

The relationship between insects and humans can be quite complex because it fits into all three categories of a biotic relationship; therefore, this association can be one of parasitism, commensalism, or mutualism. Depending on which category the relationship is at the time, people decide to take action to either combat or encourage insects in the world. If an insect has become a pest, humans seek to eliminate it from their homes; however, if the bug is helping in some way, such as controlling the population of another pest, people will promote this behavior. Hence, it can be concluded that insects have both definite advantages and disadvantages in the environment.

One problem that people have with insects is that some of them can be parasitic. Examples of parasites include bed bugs, lice, and mosquitoes. Some bugs (flies, mosquitoes, ticks, etc.) can also transmit diseases to humans. Furthermore, termites have the power to cause significant damage to structures. Another disadvantage of insects, such as locusts and weevils, is their power to destroy crops, costing the agricultural business dearly.

Although bugs have various negative impacts on humans, their positive contributions far outweigh any harm they cause. Insects are necessary for pollination, which is the most important function of their role in the environment. Wasps, bees, butterflies, and ants all contribute to the process of pollination, a mutualistic relationship between plants and insects. Pollination is vital for the production of fruits, vegetables, and nuts. This process also serves to strengthen the genetics of plants which helps overall agriculture. In the larger picture, pollination is needed to sustain all life on Earth; without it, the food chains and food webs of the world would collapse, causing the destruction of all human life.

Not only are insects advantageous because of their role in pollination, but they have many other benefits. For centuries, civilizations have valued the honey produced by bees. Other bugs create valuable goods such as silk, lacquer, beeswax, and dyes. Insects are invaluable for natural and biological control of other insect populations;

insectivorous insects consume other bugs that pose a threat to agriculture and buildings. An illustration of this relationship can be seen in the way that ladybugs hunt crop-eating aphids. Throughout history, various bugs have been celebrated in diverse cultures or relied upon for consumption. Insects are also used for the fabrication of drugs and medicinal substances. In recent years, some bugs have even aided in the studies of genetics. As a valuable agent for soil health, insects serve in the process of decomposition which helps aerate and fertilize the soil.

Despite the problems that insects can cause, they are a necessary part of life. Bugs provide many needed services and are critical for the continuity of all life. While pollination is the most significant function of insects, they also impact several other areas of life including production of various substances, biological control, culture, medicine, research, and soil health. Always keep in mind that insects are a vital part of nature that fulfill crucial functions in the environment and are beneficial to humans.

AS PESTS

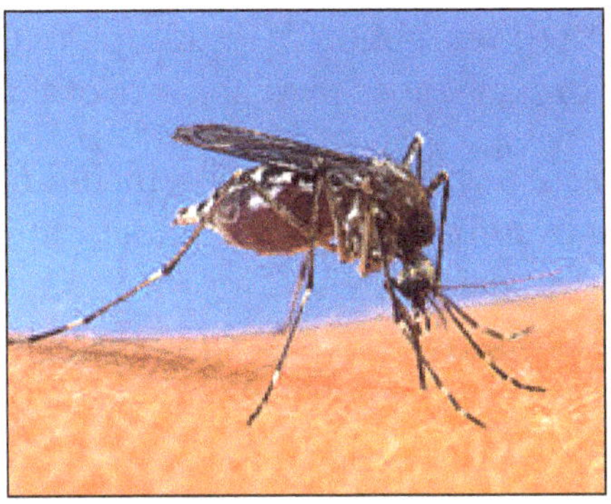

Aedes aegypti, a parasite, is the vector of dengue fever and yellow fever.

Many insects are considered pests by humans. Insects commonly regarded as pests include those that are parasitic (*e.g.* lice, bed bugs), transmit diseases (mosquitoes, flies), damage structures (termites), or destroy agricultural goods (locusts, weevils). Many entomologists are involved in various forms of pest control, as in research for companies to produce insecticides, but increasingly rely on methods of biological pest control, or biocontrol. Biocontrol uses one organism to reduce the population density of another organism—the pest—and is considered a key element of integrated pest management.

Despite the large amount of effort focused at controlling insects, human attempts to kill pests with insecticides can backfire. If used carelessly, the poison can kill all kinds of organisms in the area, including insects' natural predators, such as birds, mice and other insectivores. The effects of DDT's use exemplifies how some insecticides can threaten wildlife beyond intended populations of pest insects.

Harmful Insects

The phylloxera, a true gourmet, finds out the best vineyards and attaches itself to the best wines Cartoon from Punch.

Insects considered *pests* of some sort occur among all major living orders with the exception of Ephemeroptera (mayflies), Odonata, Plecoptera (stoneflies), Embioptera (webspinners), Trichoptera (caddisflies), Neuroptera (in the broad sense), and Mecoptera (also, the tiny groups Zoraptera, Grylloblattodea, and Mantophasmatodea). Conversely, of course, essentially all insect orders primarily have members which are beneficial, in some respects, with the exception of Phthiraptera (lice), Siphonaptera (fleas), and Strepsiptera, the three orders whose members are exclusively parasitic.

Insects are considered pests for a variety of reasons, including direct damage by feeding on crop plants in the field or by infesting stored products, indirect damage by spreading viral diseases of crop plants (especially by sucking insects such as leafhoppers), spreading disease among humans and livestock, and annoyance to humans. Examples of insects regarded as pests include the phylloxera, migratory locusts, the Colorado potato beetle, the boll weevil, Japanese beetle, aphids, mosquitoes, cockroaches, the Western corn rootworm, and some fly species.

Beneficial Insects

Many plants in the family Asteraceae attract beneficial insects.

Beneficial insects (sometimes called beneficial bugs) are any of a number of species of insects that perform valued services like pollination and pest control. The concept of *beneficial* is subjective and only arises in light of desired outcomes from a human perspective. In farming and agriculture, where the goal is to raise selected crops, insects that hinder the production process are classified as pests, while insects that assist production are considered beneficial. In horticulture and gardening; pest control, habitat integration, and 'natural vitality' aesthetics are the desired outcome with beneficial insects.

Encouraging beneficial insects, by providing suitable living conditions, is a pest control strategy, often used in organic farming, organic gardening or integrated pest management. Companies specializing in biological pest control sell many types of beneficial insects, particularly for use in enclosed areas, like greenhouses.

Types

Bee boxes at an organic farm.

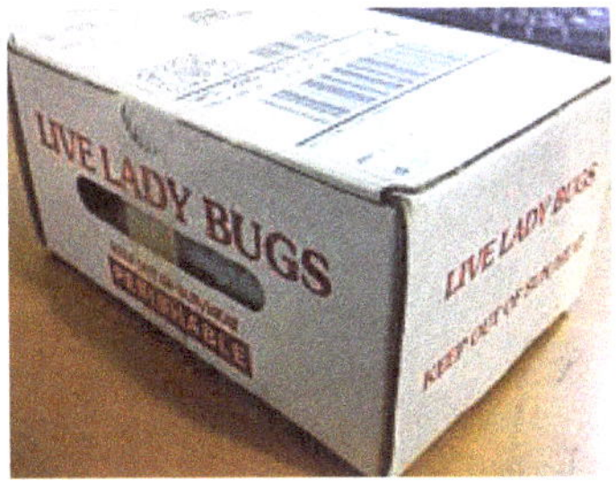

Ladybirds are a beneficial insect commonly sold for biological control of aphids.

A European mantis hunts for prey on a shrub rose.

Encarsia formosa, an endoparasitic wasp, was one of the first biological control agents developed.

Some species of bee are beneficial as pollinators, although generally only efficient at pollinating plants from the same area of origin, facilitating propagation and fruit production for many plants. Also, some bees are predatory or parasitic, killing pest insects. This group includes not only honeybees, but also many other kinds that are more efficient at pollinating. Bees can be attracted by many companion plants, especially bee balm and pineapple sage for honeybees, or Apiaceae like Queen Anne's lace and parsley, for predatory bees.

Ladybugs are generally thought of as beneficial because they eat large quantities of aphids, mites and other arthropods that feed on various plants.

Other insects commonly identified as beneficial include:

- Assassin bug,
- Damsel bug,
- Earwig,
- Green lacewing,
- Ichneumon wasp,
- Lady bugs,
- Mealybug destroyer,
- Minute pirate bug,
- Soldier beetle,
- Syrphid fly,
- Tachinid fly,
- Trichogramma wasp.

ENTOMOPHAGY

Deep-fried insects on sale at a food stall in Bangkok.

Entomophagy describes the practice of eating insects by humans (as well as by non-human species).

The eggs, larvae, pupae, and adults of certain insects have been eaten by humans from prehistoric times to the present day. Around 3,000 ethnic groups practice entomophagy. Human insect-eating is common to cultures in most parts of the world, including

Central and South America, Africa, Asia, Australia, and New Zealand. Eighty percent of the world's nations eat insects of 1,000 to 2,000 species. In some societies, primarily western nations, entomophagy is uncommon or taboo. Today, insect eating is uncommon in North America and Europe, but insects remain a popular food elsewhere, and some companies are trying to introduce insects as food into Western diets. FAO has registered some 1,900 edible insect species and estimates that there were, in 2005, some two billion insect consumers worldwide. They suggest eating insects as a possible solution to environmental degradation caused by livestock production.

Mealworms presented in a bowl for human consumption.

Entomophagy is sometimes defined broadly also to cover the eating of arthropods other than insects, including arachnids and myriapods. Insects and arachnids eaten around the world include crickets, cicadas, grasshoppers, ants, various beetle grubs (such as mealworms, the larvae of the darkling beetle), various species of caterpillar (such as bamboo worms, mopani worms, silkworms and waxworms), scorpions and tarantulas. There are over 1,900 known species of arthropods which are edible for humans.

Recent assessments of the potential of large-scale entomophagy have led some experts to suggest insects as a potential alternative protein source to conventional livestock, citing possible benefits including greater efficiency, lower resource use, increased food security, and environmental and economic sustainability.

In Non-humans

Entomophagy among animals: The giant anteater is a mammal specialized in eating insects.

Insects, nematodes and fungi that obtain their nutrition from insects are sometimes termed entomophagous, especially in the context of biological control applications. These may also be more specifically classified into predators, parasites or parasitoids, while viruses, bacteria and fungi that grow on or inside insects may also be termed "entomopathogenic".

Advantages of Eating Insects

Food Security

Deep-fried crickets.

Fried beetles in Lao cuisine.

Fried silk worm pupae sold by a street vendor in Jinan, China, one with a bite taken out of it.

The major role of entomophagy in human food security is well-documented. While more attention is needed to fully assess the potential of edible insects, they provide a natural source of essential carbohydrates, proteins, fats, minerals and vitamins, offering an opportunity to bridge the gap in protein consumption between poor and wealthy nations and also to lighten the ecological footprint. Many insects contain abundant stores of lysine, an amino acid deficient in the diets of many people who depend heavily on grain. Some argue that the combination of increasing land use pressure, climate change, and food grain shortages due to the use of corn as a biofuel feedstock will cause serious challenges for attempts to meet future protein demand.

The first publication to suggest that edible insects could ease the problems of global food shortages was by Meyer-Rochow in 1975. Insects as food and feed have emerged as an especially relevant issue in the 21st century due to the rising cost of animal protein, food and feed insecurity, environmental pressures, population growth and increasing demand for protein among the middle classes. At the 2013 International Conference on Forests for Food Security and Nutrition, the Food and Agriculture Organization of the United Nations released a publication titled *Edible insects - Future prospects for food and feed security* describing the contribution of insects to food security. It shows the many traditional and potential new uses of insects for direct human consumption and the opportunities for and constraints to farming them for food and feed. It examines the body of research on issues such as insect nutrition and food safety, the use of insects as animal feed, and the processing and preservation of insects and their products.

Small-scale Insect Farming/Minilivestock

The intentional cultivation of insects and edible arthropods for human food, referred to as "minilivestock", is now emerging in animal husbandry as an ecologically sound concept. Several analyses have found insect farming to be a more environmentally friendly alternative to traditional animal livestocking.

In Thailand, two types of edible insects (cricket and palm weevil larvae) are commonly farmed in the north and south respectively. Cricket-farming approaches throughout the northeast are similar and breeding techniques have not changed much since the technology was introduced 15 years ago. Small-scale cricket farming, involving a small number of breeding tanks, is rarely found today and most of the farms are medium- or large-scale enterprises. Community cooperatives of cricket farmers have been established to disseminate information on technical farming, marketing and business issues, particularly in northeastern and northern Thailand. Cricket farming has developed into a significant animal husbandry sector and is the main source of income for a number of farmers. In 2013, there are approximately 20,000 farms operating 217,529 rearing pens. Total production over the last six years has averaged around 7,500 tonnes per year.

In the Western world, agricultural technology companies such as Tiny Farms have been founded with the aim of modernizing insect rearing techniques, permitting the scale and efficiency gains required for insects to displace other animal proteins in the human food supply. The first domestic insect farm, LIVIN Farms Hive, has recently been successfully Kickstarted and will allow for the production of 200-500g of mealworms per week, a step toward a more distributed domestic production system.

Therapeutic Foods

In 2012, Dr. Aaron T. Dossey announced that his company, All Things Bugs, had been named a Grand Challenges Explorations winner by the Bill & Melinda Gates Foundation. Grand Challenges Explorations provides funding to individuals with ideas for new approaches to public health and development. The research project is titled "Good Bugs: Sustainable Food for Malnutrition in Children". Director of pediatric nutrition at the University of Alabama at Birmingham Frank Franklin has argued that since low calories and low protein are the main causes of death for approximately five million children annually, insect protein formulated into a ready-to-use therapeutic food similar to Nutriset's Plumpy'Nut could have potential as a relatively inexpensive solution to malnutrition. In 2009, Dr. Vercruysse from Ghent University in Belgium has proposed that insect protein can be used to generate hydrolysates, exerting both ACE inhibitory and antioxidant activity, which might be incorporated as a multifunctional ingredient into functional foods. Additionally, edible insects can provide a good source of unsaturated fats, thereby helping to reduce coronary disease.

Indigenous Cultivation

Edible insects can provide economic, nutritional, and ecological advantages to the indigenous populations that raise them. For instance, the mopane worm of South Africa provides a "flagship taxon" for the conservation of mopane woodlands. Some researchers have argued that edible insects provide a unique opportunity for insect conservation by combining issues of food security and forest conservation through a solution which includes appropriate habitat management and recognition of local traditional knowledge and enterprises. Cultures in Africa have developed unique interactions with insects as a result of their traditional ecological management practices and customs. However, senior FAO forestry officer Patrick Durst claims that "Among forest managers, there is very little knowledge or appreciation of the potential for managing and harvesting insects sustainably. On the other hand, traditional forest-dwellers and forest-dependent people often possess remarkable knowledge of the insects and their management."

Similarly, Julieta Ramos-Elorduy has stated that rural populations, who primarily "search, gather, fix, commercialize and store this important natural resource", do not exterminate the species which are valuable to their lives and livelihoods. According to the FAO, many experts see income opportunities for rural people involved in cultivation. However, adapting food technology and safety standards to insect-based foods would enhance these prospects by providing a clear legal foundation for insect-based foods.

Pest Harvesting

Larvae of the sago palm weevil, (*Rhynchophorus ferrugineus*), a serious pest of date, coconut and oil palms.

Some researchers have proposed entomophagy as a solution to policy incoherence created by traditional agriculture, by which conditions are created which favor a few insect species, which then multiply and are termed "pests". In parts of Mexico, the grasshopper *Sphenarium purpurascens* is controlled by its capture and use as food. Such strategies allow decreased use of pesticide and create a source of income for farmers totaling nearly US$3000 per family. Environmental impact aside, some argue that pesticide use is inefficient economically due to its destruction of insects which may contain up to 75 percent animal protein in order to save crops containing no more than 14 percent protein.

Environmental Benefits

The methods of matter assimilation and nutrient transport used by insects make insect cultivation a more efficient method of converting plant material into biomass than rearing traditional livestock. More than 10 times more plant material is needed to produce one kilogram of meat than one kilogram of insect biomass. The spatial usage and water requirements are only a fraction of that required to produce the same mass of food with cattle farming. Production of 150g of grasshopper meat requires very little water, while cattle requires 3290 liters to produce the same amount of beef. This indicates that lower natural resource use and ecosystem strain could be expected from insects at all levels of the supply chain. Edible insects also display much faster growth and breeding cycles than traditional livestock. An analysis of the carbon intensity of five edible insect species conducted at the University of Wageningen, Netherlands found that "the average daily gain (ADG) of the five insect species studied was 4.0-19.6 percent, the minimum value of this range being close to the 3.2% reported for pigs, whereas the maximum value was 6 times higher. Compared to cattle (0.3%), insect ADG values were much higher." Additionally, all insect species studied produced much lower amounts of ammonia than conventional livestock, though further research is needed to determine the long-term impact. The authors conclude that insects could serve as a more environmentally friendly source of dietary protein.

Economic Benefits

Insects generally have a higher food conversion efficiency than more traditional meats, measured as efficiency of conversion of ingested food, or ECI. While many insects can have an energy input to protein output ratio of around 4:1, raised livestock has a ratio closer to 54:1. This is partially due to the fact that feed first needs to be grown for most traditional livestock. Additionally, endothermic (warm-blooded) vertebrates need to use a significantly greater amount of energy just to stay warm whereas ectothermic (cold-blooded) plants or insects do not. An index which can be used as a measure is the Efficiency of conversion of ingested food to body substance: for example, only 10% of ingested food is converted to body substance by beef cattle, versus 19–31% by silkworms and 44% by German cockroaches. Studies concerning the house cricket (*Acheta domesticus*) provide further evidence for the efficiency of insects as a food source.

When reared at 30 °C or more and fed a diet of equal quality to the diet used to rear conventional livestock, crickets showed a food conversion twice as efficient as pigs and broiler chicks, four times that of sheep, and six times higher than steers when losses in carcass trim and dressing percentage are counted.

Mexican *chapulines*.

Insects reproduce at a faster rate than beef animals. A female cricket can lay from 1,200 to 1,500 eggs in three to four weeks, while for beef the ratio is four breeding animals for each market animal produced. This gives house crickets a true food conversion efficiency almost 20 times higher than beef.

Nutritional Benefits

Insects such as crickets are a complete protein (contains all nine essential amino acids) and contain a more useful amount, comparable with protein from soybeans, though less than in casein (found in foods such as cheese). They have dietary fiber and include mostly unsaturated fat and contain some vitamins and essential minerals.

Impacts of Animal Agriculture

According to the United Nations Food and Agriculture Organization (FAO), animal agriculture makes a "very substantial contribution" to climate change, air pollution, land, soil and water degradation, land use concerns, deforestation and the reduction of biodiversity. The high growth and intensity of animal agriculture has caused ecological damage worldwide; with meat production predicted to double from now to 2050, maintaining the status quo's environmental impact would demand a 50 percent reduction of impacts per unit of output. As the FAO states, animal livestock "emerges as one of the top two or three most significant contributors to the most serious environmental problems, at every scale from local to global." Some researchers argue that establishing sustainable production systems will depend upon a large-scale replacement of traditional livestock with edible insects; such a shift would require a major change in Western perceptions of edible insects, pressure to conserve remaining habitats, and an economic push for food systems that incorporate insects into the supply chain.

Greenhouse Gas Emission

In total, the emissions of the livestock sector account for 18 percent of total anthropogenic greenhouse gas emissions, a greater share than the transportation sector. Using the ratio between body growth realized and carbon production as an indicator of environmental impact, conventional agriculture practices entail substantial negative impacts as compared to entomophagy. The University of Wageningen analysis found that the CO_2 production per kilogram of mass gain for the five insect species studied was 39-129% that of pigs and 12-54% that of cattle. This finding corroborates existing literature on the higher feed conversion efficiency of insects as compared to mammalian livestock. For four of the five species studied, GHG emission was "much lower than documented for pigs when expressed per kg of mass gain and only around 1% of the GHG emission for ruminants."

Land Use

Animal livestock is the largest anthropogenic user of land. 26 percent of the Earth's ice-free terrestrial surface is occupied by grazing, while feedcrop production amounts to 33 percent of total arable land. Livestock production accounts for 70 percent of all agricultural land and 30 percent of the planet's land surface. According to the Food and Agriculture Organization, livestock activity such as overgrazing, erosion, and soil compaction, has been the primary cause of the degradation of 20 percent of the world's pastures and rangeland. Animal livestock is responsible for 64 percent of man-made ammonia emissions, which contribute significantly to acid rain. By extension, animal waste contributes to environmental pollution through nitrification and acidification of soil.

Water Pollution

According to the Food and Agriculture Organization, 64 percent of the world's population is expected to live in water-stressed basins by 2025. A reassessment of human usage and treatment of water resources will likely become necessary in order to meet growing population needs. The FAO argues that the livestock sector is a major source of water pollution and loss of freshwater resources:

> The livestock sector is probably the largest sectoral source of water pollution, contributing to eutrophication, "dead" zones in coastal areas, degradation of coral reefs, human health problems, emergence of antibiotic resistance and many others. The major sources of pollution are from animal wastes, antibiotics and hormones, chemicals from tanneries, fertilizers and pesticides used for feedcrops, and sediments from eroded pastures. Global figures are not available but in the United States, with the world's fourth largest land area, livestock are responsible for an estimated 55 percent of erosion and sediment, 37 percent of pesticide use, 50 percent of antibiotic use, and a third of the loads of nitrogen and phosphorus into freshwater resources. Livestock also affect the replen-

ishment of freshwater by compacting soil, reducing infiltration, degrading the banks of watercourses, drying up floodplains and lowering water tables.

Potential as Alternative Pet Food

There is potential for insects to be used as a protein source in insect based pet food. Novel protein sources have possible benefits for pets with sensitive gastrointestinal tracts or food allergies, as the proteins are not recognized by the animal's body, and therefore are less likely to cause irritation. Insects have also been shown to have a high palatibility to both companion and livestock animals. They have a good amino acid profile, and also contain many essential nutrients for companion animals. Insects have also been shown to have a high digestibility in pets. There have been studies done evaluating the protein quality of commonly used insects and their nutrient values in comparison to traditional pet food protein.

Disadvantages

Spoilage

Spore forming bacteria can spoil both raw and cooked insect protein, threatening to cause food poisoning. While edible insects must be processed with care, simple methods are available to prevent spoilage. Boiling before refrigeration is recommended; drying, acidification, or use in fermented foods also seem promising.

Toxicity

In general, many insects are herbivorous and less problematic than omnivores. Cooking is advisable in ideal circumstances since parasites of concern may be present. But pesticide use can make insects unsuitable for human consumption. Herbicides can accumulate in insects through bioaccumulation. For example, when locust outbreaks are treated by spraying, people can no longer eat them. This may pose a problem since edible plants have been consumed by the locusts themselves.

In some cases, insects may be edible regardless of their toxicity. In the Carnia region of Italy, moths of the Zygaenidae family have been eaten by children despite their potential toxicity. The moths are known to produce hydrogen cyanide precursors in both larvae and adults. However, the crops of the adult moths contain cyanogenic chemicals in extremely low quantities along with high concentrations of sugar, making *Zygaena* a convenient supplementary source of sugar during the early summer. The moths are very common and easy to catch by hand, and the low cyanogenic content makes *Zygaena* a minimally risky seasonal delicacy.

Cases of lead poisoning after consumption of chapulines were reported by the California Department of Health Services in November 2003. Adverse allergic reactions are also a possible hazard.

Edible Insects

Frequently Consumed Insect Species

Estimates of numbers of edible insect species consumed globally range from 1,000 to 2,000. These species include 235 butterflies and moths, 344 beetles, 313 ants, bees and wasps, 239 grasshoppers, crickets and cockroaches, 39 termites, and 20 dragonflies, as well as cicadas. Which species are consumed varies by region due to differences in environment, ecosystems, and climate.

The table below lists the top five insect orders consumed by humans worldwide, retrieved from Edible Insects: Future Prospects for Food and Feed Security by Arnold van Huis, Joost Van Itterbeeck, Harmke Klunder, Esther Mertens, Afton Halloran, Giulia Muir and Paul Vantomme.

Order of insect	Common name	Consumption rate worldwide by human population (%)
Coleoptera	Beetles	31
Lepidoptera	Butterflies, moths	18
Hymenoptera	Bees, wasps, ants	14
Orthoptera	Grasshoppers, locusts, crickets	13
Hemiptera	Cicadas, leafhoppers, planthoppers	10

For a list of edible insects consumed locally see: List of edible insects by country.

Edible Insects for Industrialized Mass Production

Freeze-dried house crickets as food (or food ingredient).

Mealworms as food (or food ingredient).

Buffalo worms as food (or ingredient).

In Western markets such as Europe and North America, academics as well as large-scale insect food producers such as Entomofarms in Canada, Aspire Food Group in the United States, Protifarm in the Netherlands, and Bühler Group in Switzerland, focus on four insects species suitable for human consumption as well as industrialized mass production:

- House cricket (Acheta domesticus).
- European migratory locust (Locusta migratoria).

- Mealworms (Tenebrio molitor) as larvae.
- Lesser mealworms (Alphitobius diaperinus) as larvae, mostly marketed under the term buffalo worms.

Nutritional Profile

Insects are nutrient efficient compared to other meat sources. Insects such as crickets are a complete protein and contain a useful amount, comparable with protein from soybeans, though less than in casein (found in foods such as cheese). They have dietary fiber and include mostly unsaturated fat and contain some vitamins, such as vitamin B12, riboflavin and vitamin A, and essential minerals.

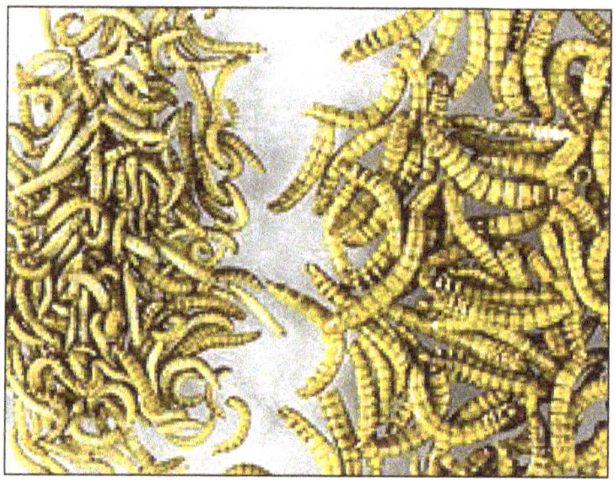

Freeze-dried mealworms and buffalo worms (lesser mealworm).

Locusts contain between 8 and 20 milligrams of iron for every 100 grams of raw locust. Beef on the other hand contains roughly 6 milligrams of iron in the same amount of meat. Crickets as well are very efficient in terms of nutrients. For every 100 grams of substance crickets contain 12.9 grams of protein, 121 calories, and 5.5 grams of fat. Beef contains more protein containing 23.5 grams in 100 grams of substance, but also has roughly 3 times the calories, and four times the amount of fat as crickets do in 100 grams. So, per 100 grams of substance, crickets contain only half the nutrients of beef, except for iron. High levels of iron are implicated in bowel cancer and heart disease.

Nutritional value per 100 g	Mealworms (Tenebrio molitor)	Buffalo worms (Alphitobius diaperinus)	House crickets (Acheta domesticus)	Migratory locust (Locusta migratoria)
Energy	550 kcal / 2303 KJ	484 kcal/ 2027 KJ	458 kcal/ 1918 KJ	559 kcal/ 2341 KJ
Fat Of which saturated fatty acids	37,2 g 9 g	24,7 g 8 g	18,5 g 7 g	38,1 g 13,1 g
Carbohydrates Of which sugars	5,4 g 0 g	6,7 g 0 g	0 g 0 g	1,1 g 0 g
Protein	45,1 g	56,2 g	69,1 g	48,2 g
Salt	0,37 g	0,38 g	1,03 g	0,43 g

Farming, Production and Processing

Edible insects are raised as livestock in specialized insect farms. In North American as well as European countries such as the Netherlands or Belgium, insects are produced under strict food law and hygiene standard for human consumption.

Several variables apply, such as temperature, humidity, feed, water sources, housing, depending on the insect species. The insects are raised from eggs to larvae status (mealworms, lesser mealworms) or to their mature form (crickets, locusts), and then killed, in industrialized insect farms by lowering the temperature. After that the insects are freeze-dried and packed whole, or pulverized to insect powder (insect flour), to be processed in other food products such as bakery products, or snacks.

Crickets being raised for human consumption.

Aside from nutritional composition and digestibility, insects are also selected for ease of rearing by the producer. This includes susceptibility to disease, efficiency of feed conversion, developmental rate and generational turnover.

Insect Food Products

The following processed food products are produced by several producers in North America, Canada, and the EU:

- Insect flour: Pulverized, freeze-dried insects (e.g., cricket flour).
- Insect burger: Hamburger patties made from insect powder/insect flour (mainly from worms or from house cricket) and further ingredients.
- Insect fitness bars: Protein bars containing insect powder (mostly house crickets).
- Insect pasta: Pasta made of wheat flour, fortified with insect flour (house crickets or mealworms).
- Insect bread (Finnish Sirkkaleipä): Bread baked with insect flour (mostly house crickets).
- Insect snacks (Finnish BugBites): Oat snack with cultivated house cricket powder.

Insect burger made by Essento containing mealworms.

Fusilli made of cricket flour.

INSECTS IN MEDICINE

Insects have long been used in medicine, both traditional and modern, sometimes with little evidence of their effectiveness. For the purpose of the article, and in line with custom, medicinal uses of other arthropods such as spiders are included.

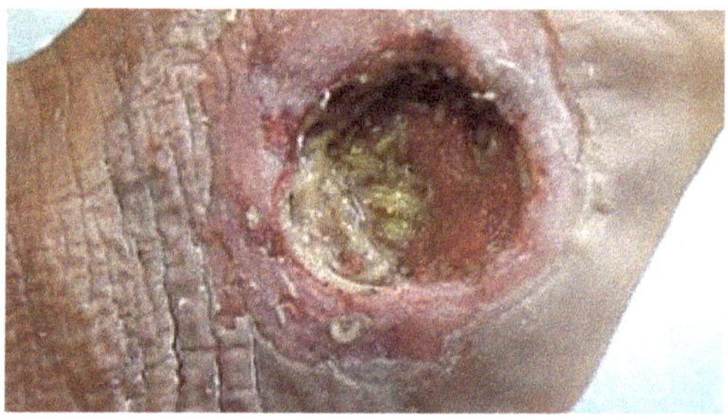

Maggot debridement therapy on a wound on a diabetic foot.

Traditional and Alternative Uses

The medicinal uses of insects and other arthropods worldwide have been reviewed by Meyer-Rochow, who provides examples of all major insect groups,spiders, worms and molluscs and discusses their potential as suppliers of bioactive components. Using insects (and spiders) to treat various maladies and injuries has a long tradition and, having stood the test of time, can be effective and provide results. However, sometimes folk-medicinal "logic" was based on the Doctrine of Signatures = "let likes be cured by likes"and had, if any at all, little more than a psychological effect. For example, to treat cases of constipation, dung beetles were prescribed; to slim down stick insects were thought to help; hairy tarantulas seemed the right treatment for hair loss and fat grubs resembling the swollen limb caused by the parasite Wucheria bancrofti were expected to help the elephantiasis sufferer. An organism bearing parts that resemble human body parts, animals, or other objects, was thought to have useful relevance to those parts, animals or objects. So, for example, the femurs of grasshoppers, which were said to resemble the human liver, were used to treat liver ailments by the indigenous peoples of Mexico. This doctrine is common throughout traditional and alternative medicine, but is most prominent where medical traditions are broadly accepted, as in traditional Chinese medicine and Ayurveda, and less by community and family based medicine, as is more common in parts of Africa.

Traditional Chinese Medicine

Traditional Chinese medicine includes the use of herbal medicine, acupuncture, massage, exercise, and dietary therapy. It is a typical component of modern medical care throughout East Asia and in some parts of Southeast Asia (such as Thailand). Insects are very commonly incorporated as part of the herbal medicine component of traditional Chinese medicine, and their medical properties and applications are broadly accepted and agreed upon. Some brief examples follow:

The Chinese Black Mountain Ant, Polyrhachis vicina, is supposed to act as a cure all and is widely used, especially by the elderly. It is said to prolong life, to have anti-aging properties, to replenish Qi, and to increase virility and fertility. Recent interest in the ants' medicinal qualities has led to British researchers to study the extract's potential to serve as a cancer-fighting agent. Chinese Black Mountain Ant extract is typically consumed mixed with wine.

India and Ayurveda

Ayurveda is ancient traditional Indian treatment almost universally incorporated alongside Western medicine as a typical component of medical treatment in India. Although Ayurvedic medicine is often effective, doses can be inconsistent, and may sometimes be contaminated with toxic heavy metals. Some brief examples to follow:

Termite is said to cure a variety of diseases, both specific and vague. Typically the mound or a portion of the mound is dug up and the termites and the architectural components of the mound are together ground into a paste which is then applied topically to the affected areas or, more rarely, mixed with water and consumed. This treatment was said to cure ulcers, rheumatic diseases, and anemia. It was also suggested to be a general pain reliever and health improver.

The Jatropha Leaf Miner, a lepidopteran which feeds preferentially on Jatropha, is an example of a major insect agricultural pest which is also a medicinal remedy. The larvae, which are also the form of the insect with the greatest economic impact on agriculture, are harvested, boiled, and mashed into a paste which is administered topically and is said to induce lactation, reduce fever, and soothe gastrointestinal tracts.

Africa

Unlike China and India, the traditional insect medicine of Africa is extremely variable. It is largely regional, with few, if any, major agreements on which insects are useful as treatments for which ailments. Most insect medicinal treatments are passed on through communities and families, rather than being taught in university settings, as Traditional Chinese Medicine and Ayurveda sometimes are; furthermore, most traditional medicine practices necessitate a person in a "healer" role. Some brief examples to follow:

Grasshopper is both commonly eaten as a delicacy and an excellent source of protein and is consumed for medicinal purposes. These insects are typically collected, dried in the sun, and then ground into a powder. The powder can then be turned into a paste when mixed with water and ash and applied to the forehead to alleviate the pain of violent headaches. Additionally, the headaches themselves can be prevented by a "healer" inserting the paste under the skin at the nape of the afflicted person's neck.

Termites are also used in parts of Africa much like they are in India. Parts of the mound are dug up, boiled, and turned into a paste, which can then be applied to external wounds to prevent infection or consumed to treat internal hemorrhages. termites are used not only as a form of medicine, but also as a medical device. If a "healer" wants to insert a medicine subcutaneously, they will often spread that medicine on the skin of the patient, and then agitate a termite and place the insect on the skin of the patient. When the termite bites, its mandibles effectively serve as an injection device.

Americas

The Americas were more highly influenced by the Doctrine of Signatures than China, India, or Africa, most likely because of their colonial history with Europe. The majority of insect use in medicine is associated with Central America and parts of South America, rather than North America, and most of it is based on the medical techniques of indigenous peoples. Currently, insect medicine is practiced much more rarely than in China, India, or Africa, though it is still relatively common in rural areas with large indigenous populations. Some examples to follow:

Chapulines, or grasshoppers, are commonly consumed as a toasted regional dish in some parts of Mexico, but they are also used medicinally. They are said to serve as diuretic to treat kidney diseases, to reduce swelling, and to relieve the pain of intestinal disorders when they are consumed. However, there are some risks associated with consuming chapulines, as they are known to harbor nematodes which may be transmitted to humans upon consumption.

Much like the termites of Africa, ants were sometimes used as medicinal devices by the indigenous peoples of Central America. The soldier cast of the Army ant would be collected and used as living sutures by Mayans. This involved agitating an ant and holding its mandibles up to the wound edges; when it bit down, the thorax and abdomen were removed, leaving the head holding the wound together. The ant's salivary gland secretions were reputed to have antibiotic properties. The venom of the Red harvester ant was used to treat rheumatism, arthritis, and poliomyelitis via the immunological reaction produced by its sting. This technique, in which ants are allowed to sting afflicted areas in a controlled manner, is still used in some arid rural areas of Mexico.

The silkworm, Bombyx mori, was also commonly consumed both as a regional food and for medicinal purposes in Central America after it was brought to the New World by the Spanish and Portuguese. Only the immatures are consumed. Boiled pupae were eaten to treat apoplexy, aphasy, bronchitis, pneumonia, convulsions, hemorrhages, and frequent urination. The excrement produced by the larvae is also eaten to improve circulation and alleviate the symptoms of cholera (intense vomiting and diarrhea).

Honey Bee Products

Honey bee products are used medicinally across Asia, Europe, Africa, Australia, and the Americas, despite the fact that the honey bee was not introduced to the Americas until the colonization by Spain and Portugal. They are by far the most common medical insect product, both historically and currently.

Honey is the most frequently referenced medical bee material. It can be applied to skin to treat excessive scar tissue, rashes, and burns, and can be applied as a poultice to eyes to treat infection. It is also consumed for digestive problems and as a general health restorative, and can be heated and consumed to treat head colds, cough, throat infections, laryngitis, tuberculosis, and lung diseases.

Additionally, apitoxin, or honey bee venom, can be applied via direct stings to relieve arthritis, rheumatism, polyneuritis, and asthma. Propolis, a resinous, waxy mixture collected by honeybees and used as a hive insulator and sealant, is often consumed by menopausal women because of its high hormone content, and it is said to have antibiotic, anesthetic, and anti-inflammatory properties. Royal jelly is used to treat anemia, gastrointestinal ulcers, arteriosclerosis, hypo- and hypertension, and inhibition of sexual libido. Finally Bee bread, or bee pollen, is eaten as a generally health restorative, and is said to help treat both internal and external infections. All of these honey bee products are regularly produced and sold, especially online and in health food stores, though none are yet approved by the FDA.

Modern Scientific Uses

Though insects were widely used throughout history for medical treatment on nearly every continent, relatively little medical entomological research has been conducted since the revolutionary advent of antibiotics. Heavy reliance on antibiotics, coupled with discomfort with insects in Western culture limited the field of insect pharmacology until the rise of antibiotic resistant infections sparked pharmaceutical research to explore new resources. Arthropods represent a rich and largely unexplored source of new medicinal compounds.

Maggot Therapy

Maggot therapy is the intentional introduction of live, disinfected blow fly larvae (maggots) into soft tissue wounds to selectively clean out the necrotic tissue. This helps to prevent infection; it also speeds healing of chronically infected wounds and ulcers. Military surgeons since classical antiquity noticed that wounds which had been left untreated for several days, and which had become infested with maggots, healed better than wounds not so infested. Maggots secrete several chemicals that kill microbes, including allantoin, urea, phenylacetic acid, phenylacetaldehyde, calcium carbonate, proteolytic enzymes, and many others.

Maggots were used for wound healing by the Maya and by indigenous Australians. More recently, they were used in Renaissance Europe, in the Napoleonic Wars, the American Civil War, and in the First and Second World Wars. It continues to be used in military medicine.

Apitherapy

Apitherapy is the medical use of honeybee products such as honey, pollen, bee bread, propolis, royal jelly and bee venom. One of the major peptides in bee venom, called Melittin, has the potential to treat inflammation in sufferers of Rheumatoid arthritis and Multiple sclerosis. Melittin blocks the expression of inflammatory genes, thus reducing swelling and pain. It is administered by direct insect sting, or intramuscular injections. Bee products demonstrate a wide array of antimicrobial factors and in laboratory studies and have been shown to kill antibiotic resistant bacteria, pancreatic cancer cells, and many other infectious microbes.

Blister Beetle and Spanish Fly

Spanish fly is an emerald-green beetle, Lytta vesicatoria, in the blister beetle family (Meloidae). It and other such species were used in preparations offered by traditional apothecaries. The insect is the source of the terpenoid cantharidin, a toxic blistering agent once used as an aphrodisiac.

Blood-feeding Insects

Many blood-feeding insects like ticks, horseflies, and mosquitoes inject multiple bioactive compounds into their prey. These insects have been used by practitioners of Eastern Medicine for hundreds of years to prevent blood clot formation or thrombosis. However, modern medical research has only recently begun to investigate the drug development potential of blood-feeding insect saliva. These compounds in the saliva of blood feeding insects are capable of increasing the ease of blood feeding by preventing coagulation of platelets around the wound and provide protection against the host's immune response. Currently, over 1280 different protein families have been associated with the saliva of blood feeding organisms. This diverse range of compounds may include:

- Inhibitors of platelet aggregation, ADP, arachidonic acid, thrombin, and PAF,
- Anticoagulants,
- Vasodilators,
- Vasoconstrictors,
- Antihistamines,
- Sodium channel blockers,
- Complement inhibitors,
- Pore formers,
- Inhibitors of angiogenesis,
- Anaesthetics,
- Amps and microbial pattern recognition molecules,
- Parasite enhancers/activators.

Currently, some preliminary progress has been made with investigation of the therapeutic properties of tick anticoagulant peptide (TAP) and Ixolaris a novel recombinant tissue factor pathway inhibitor (TFPI) from the salivary gland of the tick, Ixodes scapularis. Additionally, Ixolaris, a tissue factor inhibitor has been shown to block primary tumor growth and angiogenesis in a glioblastoma model. Despite the strong potential of these compounds for use as anticoagulants or immunomodulating drugs no modern medicines, developed from the saliva of blood-sucking insects, are currently on the market.

Arachnids

Like plants and insects, arachnids have also been used for thousands of years in traditional medical practices. Recent scientific research in natural bioactive factors has increased, leading to a renewed interest in venom components in many animals. In 1993 Margatoxin was synthesized from the venom of the Centruroides margaritatus the Central American bark scorpion. It is a peptide that selectively inhibits voltage-dependent potassium channels. Patented by Merck, it has the potential to prevent neointimal hyperplasia, a common cause of bypass graft failure.

In addition to medical uses of arachnid defense compounds, a great amount of research has recently been directed toward the synthesis and use of spider silk as a scaffolding for ligament generation. Spider silk is an ideal material for the synthesis of medical skin grafts or ligament implants because it is one of the strongest known natural fibers and triggers little immune response in animals. Spider silk may also be used to make fine sutures for stitching nerves or eyes to heal with little scarring. Medical uses of spider silk are not a new idea. Spider silks have been used for thousands of years to fight infection and heal wounds. Efforts to produce industrial quantities and qualities of spider silk in transgenic goat milk are underway.

Psychoactive Scorpions

Recent news reports claim that use of scorpions for psychoactive purposes is gaining in popularity in Asia. Heroin addicts in Afghanistan are purported to smoke dried scorpions or use scorpion stings to get high when heroin is not available. The 'scorpion sting craze' has also increased in India with a decreasing availability of other drugs and alcohol available to youth. Young people are reportedly flocking to highway sides where they can purchase scorpion stings that after several minutes of intense pain, supposedly produce a six- to eight-hour feeling of wellbeing.

IN FORENSIC INVESTIGATIONS

Forensic entomology is the study of the application of insects and other arthropods in criminal investigation. Insects or arthropods are found in a decomposing vertebrate corpse or carrion. These insect colonizers can be used to estimate the time of death i.e., time interval between death and corpse discovery, also called postmortem index (PMI), movement of the corpse, manner and cause of death and association of suspects at the death scene.

Arthropods and their Association with Postmortem Changes of the Human Body

As soon as death occurs, cells start dying and enzymes start digesting the cells inside out in a process called autolysis. The body starts decomposing. Bacteria present in the gastrointestinal tract start destroying the soft tissue producing liquids and gases like hydrogen sulphide, carbon dioxide, methane, ammonia, sulfur dioxide and hydrogen. The volatile molecules called apeneumones escaping from the decomposing body attract insects. Researchers are able to isolate the volatile chemicals released at different stages of decomposition of the body. The volatile molecules released during each stage can modify the insect behavior. Based on the studies done by Crag *et al.*, in 1950 it was found that putrative sulfur-based compounds were responsible for initially attracting the flies to the decomposing carcass but egg laying or oviposition of the flies are induced by ammonium-rich compounds present on the carrion.

According to Smith four categories of insects can be found on decomposing carrion: i) Necrophagous species feeding on the carrion; ii) Predators and parasites feeding on the necrophagous species: this group also contains schizophagous species which feed on the body first and which become predaceous on the later stages; iii) Omnivorous species feeding on the carrion and other arthropods like ants, wasps and some beetles; iv) Other species like springtails and spiders which use the corpse as an extension of their environment. The first two groups are found to be more important for the purpose of forensic entomology. They are mainly from the species of the order Diptera (flies) and Coleoptera (beetles). The succession waves in which the arthropods colonize the carrion depends on the state of decomposition of the carrion.

Insects mostly involved in the forensic investigations are true flies or Diptera. The predominant species in this order are Calliphoridae (blow flies), Sacrophagidae (flesh flies) and Muscidae (house flies). Calliphoridae (blow flies), Sacrophagidae (flesh flies) may arrive within minutes following death. Muscidae (house flies) delay colonization until the body reaches bloat stages of decomposition. Calliphoridae adults are commonly shiny with metallic coloring, often with blue, green, or black thoraxes and abdomen. Sarcophagidae are medium-sized flies with black and gray longitudinal stripes on the thorax and checkering on the abdomen. The adult Muscidae are 8-12 mm long. Their thorax is gray, with four longitudinal dark lines on the back. The whole body is covered with hair-like projections. Usual areas of oviposition or egg laying are the natural body openings and wounds. When they hatch, they produce a larva called maggot. They are small peg-shaped organisms with a pair of mouth hooks on the anterior end for feeding. Maggots grow rapidly passing through the three stages or instars, reaching the full size. Once the full size is reached feeding stops and they migrate to drier areas and they begin pupariation (pupa formation). At this stage the outer skin of the maggot becomes hardened and forms a protective encasement eventually emerging as a fly.

According to the studies done by K. Tullies and M. L Goff on exposed carrion in a tropical rainforest, it was found that the decomposition process was best divided into five stages on the basis of physical appearance of carcasses, internal temperatures and characteristic insect populations:

- Fresh stage (Days 1-2): Which begins at the moment of death and ends when the bloating of the carcass is observed. Even though autolysis occurs at this stage gross morphological changes do not occur at this point. The estimation of the time of death by entomological data after 24 hrs is more accurate than medical examiner's estimation based on the soft tissue examination. Insects were seen attracted within the first 10 min of death to the carcass but no egg laying (oviposition) was found during this state. Cellular breakdown occurs during this stage without morphologic alterations. Even though morphological changes and odors are not obvious to humans, the chemicals released from the cellular breakdown attracts insects even in this early stage.

- Bloated stage (Days 2-7): Putrefaction begins at this stage. Gases produced by the metabolic activities of anaerobic bacteria cause an inflation of the abdomen and the carcass forming a balloon-like appearance during the later part. Arthropod activities combined with the putrefaction processes cause internal temperatures of the carcass to rise. The greatest numbers of adult Diptera were attracted to the carcasses during this stage. By the fourth day, first- and early second-instar or larval stages Diptera were present. By the beginning of Day 2, several predators of Diptera larvae were also recovered from the carcasses.

- Decay stage (Days 5-13): Abdominal wall is penetrated, resulting in the deflation of the carcass and ending the bloated stage, the internal temperature rises to 14 degrees above the ambient temperature followed by a drop signifying the end of the decay stage. Decaying odors are high during increased temperatures and drop with a fall in temperature. There is a steady decrease in the weight of the carcass by 10th day. There is a conversion of carcass biomass to dipteran larval biomass. The larvae subsequently depart from the carcass to pupate.

- Post-decay stage (Days 10-23): The post-decay stage begins when most of the Diptera larvae leave the carcass, leaving behind bones, cartilage, hair, small portions of tissue, and a large amount of wet, viscous material known as by-products of decay (BOD). The BOD is the major site of arthropod activity during this stage.

- Remains stage (Days 18-90+): This stage is characterized by bones with little cartilage remaining and the BOD has dried up. The transition from post-decay to remains stage is gradual, with declining adult and larval Diptera populations.

Steps in Estimating the Postmortem Index with Insect Larvae

The insect larvae present on the dead body can provide evidence for the estimation of PMI up to one month. Correct species identification is the initial step. Different species differ in their growth rates and maturation. For estimating the PMI, age of the larvae has to be determined. By measuring the length or dry weight of the oldest larvae and comparing it with the reference data, age of the larvae can be estimated. The rate of development of the larvae is dependent on the surrounding ambient temperatures. Each stage of development has its temperature requirement hence each species has its own defined number of accumulated degree days or accumulated degree hours to complete its development. Once the thermal history of the larvae is obtained, it can be compared with temperatures at the death scene and PMI can be estimated. The first-generation adult flies can also be used to determine the age. They can be identified by the shriveled wings, and tiny abdomen with dull grey colour. When insects colonizing the carrion in a particular area is known, an insect colonizing succession model can also be used to estimate the PMI.

Using Insect Data for Determining the Site of Crime

There are reported differences in the species of insects involved with the decomposing corpse in different habitats and environments. A careful examination can reveal species variation, as species associated with one type of habitat present on a corpse is found to be different from those when the corpse is transported after death.

DNA Analysis for Species Identification

Identifying the correct species is the important initial step for estimating the age of the larvae. Morphological comparison is usually used for species identification which requires special expertise and is often time-consuming. In order to overcome this difficulty species identification can be done by polymerized chain reaction amplification suitable regions of the larvae genomes and comparing it with reference data.

Entomotoxicology

The larvae of the flies which feed on the carrion can accumulate drugs ingested by the deceased person. Bodies which are in advanced stages of decomposition or which are skeletonized are difficult to examine for toxicological substances. In these instances, the larvae feeding on this body can be macerated and analyzed with techniques like thin-layer chromatography, gas chromatography and/or mass spectrometry. Toxins can influence the stages of development of the larvae. Cocaine and heroin in the carcass can accelerate the larval development. Poisons like malthione in the carrion can delay the insect colonization.

References

- The-relationship-between-insects-and-humans: pestops.com, Retrieved 05 January, 2020
- Nakamaru, M; Iwasab, Y; Nakanishic, J (October 2003). “Extinction risk to bird populations caused by DDT exposure”. Chemosphere. 53 (4): 377–387. Bibcode:2003Chmsp..53..377N. doi:10.1016/S0045-6535(03)00010-9. PMID 12946395
- Ramos-Elorduy, Julieta (2009). “Anthropo-Entomophagy: Cultures, Evolution And Sustainability”. Entomological Research. 39 (5): 271–288. doi:10.1111/j.1748-5967.2009.00238.x
- Forensic entomology. Naturwissenschaften. 2004;91:51–65

PERMISSIONS

All chapters in this book are published with permission under the Creative Commons Attribution Share Alike License or equivalent. Every chapter published in this book has been scrutinized by our experts. Their significance has been extensively debated. The topics covered herein carry significant information for a comprehensive understanding. They may even be implemented as practical applications or may be referred to as a beginning point for further studies.

We would like to thank the editorial team for lending their expertise to make the book truly unique. They have played a crucial role in the development of this book. Without their invaluable contributions this book wouldn't have been possible. They have made vital efforts to compile up to date information on the varied aspects of this subject to make this book a valuable addition to the collection of many professionals and students.

This book was conceptualized with the vision of imparting up-to-date and integrated information in this field. To ensure the same, a matchless editorial board was set up. Every individual on the board went through rigorous rounds of assessment to prove their worth. After which they invested a large part of their time researching and compiling the most relevant data for our readers.

The editorial board has been involved in producing this book since its inception. They have spent rigorous hours researching and exploring the diverse topics which have resulted in the successful publishing of this book. They have passed on their knowledge of decades through this book. To expedite this challenging task, the publisher supported the team at every step. A small team of assistant editors was also appointed to further simplify the editing procedure and attain best results for the readers.

Apart from the editorial board, the designing team has also invested a significant amount of their time in understanding the subject and creating the most relevant covers. They scrutinized every image to scout for the most suitable representation of the subject and create an appropriate cover for the book.

The publishing team has been an ardent support to the editorial, designing and production team. Their endless efforts to recruit the best for this project, has resulted in the accomplishment of this book. They are a veteran in the field of academics and their pool of knowledge is as vast as their experience in printing. Their expertise and guidance has proved useful at every step. Their uncompromising quality standards have made this book an exceptional effort. Their encouragement from time to time has been an inspiration for everyone.

The publisher and the editorial board hope that this book will prove to be a valuable piece of knowledge for students, practitioners and scholars across the globe.

The publisher and the editorial board hope that this book will prove to be a valuable piece of knowledge for researchers, students, practitioners and scholars across the globe.

INDEX

www.ingramcontent.com/pod-product-compliance
Lightning Source LLC
LaVergne TN
LVHW080613210926
848485LV00003B/151
9781639873302